# CONTENTS

## RAND
**Arroyo Center**

# *Identifying Potential*
# *Ethnic Conflict*

## Application of
## a Process Model

Edited by
**Thomas S. Szayna**

Prepared for the United States Army

For more information on the RAND Arroyo Center, contact the Director of Operations, (310) 393-0411, extension 6500, or visit the Arroyo Center's Web site at http://www.rand.org/organization/ard/

This report outlines a model for anticipating the occurrence of communitarian and ethnic conflict. The intended audience for this report is the intelligence community, though analysts and scholars involved in conflict prevention also should find it useful.

The model is not a mechanistic tool, but a process-based heuristic device with a threefold purpose: (1) to order the analyst's thinking about the logic and dynamics of potential ethnically based violence and to aid in defining the information-collection requirements of such an analysis; (2) to provide a general conceptual framework about how ethnic grievances form and group mobilization occurs and how these could lead to violence under certain conditions; and (3) to assist the intelligence community with the long-range assessment of possible ethnic strife. The framework presented here is not meant to substitute for the knowledge, reasoning, or judgment of intelligence analysts. It is simply a tool to help order and organize the information and identify information gaps.

This report is the final product of a multiyear project entitled "Ethnic Conflict and the Processes of State Breakdown." The project included two stages: model design and development, and model testing and validation. The project aimed to improve the Army's ability to anticipate communitarian and ethnic conflict as one aspect of its overall strategic planning and threat assessment. This report includes a revised version of most of the initial report on the first part of the project (Ashley J. Tellis, Thomas S. Szayna, and James A. Winnefeld, *Anticipating Ethnic Conflict*, Santa Monica, CA: RAND, MR-857-A, 1997).

The research was sponsored by the Deputy Chief of Staff for Intelligence, U.S. Army, and was conducted in the Strategy, Doctrine, and Resources Program of the RAND Arroyo Center. The Arroyo Center is a federally funded research and development center sponsored by the United States Army.

For comments and further information, please contact the editor at Tom_Szayna@rand.org.

# FIGURES

# TABLES

## Matrixes

# ACKNOWLEDGMENTS

The editor and the authors are grateful to LTG Claudia Kennedy, USA, the Army's Deputy Chief of Staff for Intelligence (DCSINT), for sponsoring the research. We also thank COL William Speer, USA, Ken Knight, Patrick Neary, and Eric Kraemer (all previously or currently at DCSINT) for their support and understanding throughout the project. We also thank James Quinlivan, Thomas McNaugher, Richard Darilek, and David Kassing at RAND for providing comments in the course of the work and for their support of the research.

James Winnefeld, in addition to playing a critical role in the development of the model, reviewed thoroughly all of the case studies and made helpful suggestions on the whole model-validation process. Robert Dorff reviewed carefully the entire document. Donald Horowitz, Daniel Byman, Marten van Heuven, Kotera Bhimaya, and Gojko Vuckovic also provided constructive comments on the model and/or the individual cases of application of the model. We also thank the many analysts in the intelligence community who provided comments and support in the course of the project.

Finally, Donna Keyser worked hard to improve the format of the case studies and Nikki Shacklett had the daunting task of editing the whole volume.

# Chapter One

# INTRODUCTION

## Thomas S. Szayna and Ashley J. Tellis

## THE CONTEXT

Intrastate communitarian strife, often dubbed "ethnic conflict," has gained much attention in the aftermath of the Cold War. Certainly, intrastate conflict has been by far the dominant form of strife in the world in the 1990s. Of the 108 armed conflicts that took place in the world in the period 1989–1998, only seven were interstate wars.[1] Most of the intrastate conflicts have had a communitarian aspect. Several of the communitarian conflicts have led to tens and even hundreds of thousands of casualties. Despite its higher prominence, the phenomenon of communitarian conflict is not new. Occasionally, major world events, such as decolonization, the collapse of empires, and, more recently, the collapse of communism, have caused spikes in the incidence of communitarian conflict.[2] But intrastate communitarian conflicts have been with us throughout the 20th century,[3] and growing social pressures associated with moderniza-

---

[1]Peter Wallensteen and Margareta Sollenberg, "Armed Conflict, 1989–98," *Journal of Peace Research*, Vol. 36, No. 5, 1999, pp. 593–606. Armed conflict is defined as "a contested incompatibility which concerns government and/or territory where the use of armed force between two parties, of which at least one is the government of a state, results in at least 25 battle-related deaths" (p. 605).

[2]After steadily rising from 47 armed conflicts in 1989 to 55 in 1992, the number and intensity of armed strife in the world declined steadily to 35 in 1995 and, since that time, has remained near that level (36 in 1998). Wallensteen and Sollenberg, p. 594.

[3]See the appendix in Roy Licklider, "The Consequences of Negotiated Settlements in Civil Wars, 1945–1993," *American Political Science Review*, Vol. 89, No. 3, September 1995, pp. 681–690.

tion suggest that such conflicts will remain with us for a long time. Such strife may even increase in the years ahead.

What is new, however, is that since the end of the Cold War, the U.S. armed forces, and the Army specifically, routinely have been called upon to conduct peace and stability operations aimed at preventing, quelling, or dealing with the consequences of communitarian strife. Moreover, the post–Cold War peace operations are different from earlier peace operations in size, scope, and complexity. Rather than stemming from a purposive grand strategy, U.S. participation in such peace operations stems from its position of leadership in the world, humanitarian considerations, and a region-specific combination of U.S. incentives and constraints.[4]

Facing the serious prospect of further involvement in peace operations in the years ahead, the Army has had to grapple with the problem of what that implies for its readiness, training, equipping, doctrine, and deployment.[5] While the primary mission of the Army and the U.S. armed forces in general will remain the fighting of wars and protecting U.S. interests in the world, peace operations (spanning the spectrum from traditional peacekeeping and humanitarian intervention to postconflict stability and peace enforcement) will continue to place severe demands on the U.S. armed forces for the foreseeable future, with the Army (and the Marine Corps) most affected. To put it bluntly, there "will be more Somalias, Rwandas, Haitis and Burundis in the future,"[6] and the Army will be called upon to deal

---

[4]There is a voluminous literature on the rationale for U.S. participation in peace operations in the 1990s. For an insightful look at the logic behind U.S. choices of intervening and not intervening, see Benjamin Miller, "The Logic of U.S. Military Interventions in the Post–Cold War Era," *Contemporary Security Policy*, Vol. 19, No. 3, December 1998, pp. 72–109.

[5]For a survey of these issues, see James A. Winnefeld, Margaret C. Harrell, Robert D. Howe, Arnold Kanter, Brian Nichiporuk, Paul S. Steinberg, Thomas S. Szayna, and Ashley J. Tellis, *Intervention in Intrastate Conflict: Implications for the Army in the Post–Cold War Era*, Santa Monica, CA: RAND, MR-554/1-A, 1995; William T. Johnsen, *Pandora's Box Reopened: Ethnic Conflict in Europe and Its Implications*, U.S. Army War College, Strategic Studies Institute, December 1994; and Major General William A. Stofft and Gary L. Guertner, *Ethnic Conflict: Implications for the Army of the Future*, U.S. Army War College, Strategic Studies Institute, March 1994.

[6]Walter Clarke and Robert Gosende, "Keeping the Mission Focused: The Intelligence Component in Peace Operations," *Defense Intelligence Journal*, Vol. 5, No. 2, Fall 1996, pp. 47–69.

with some of them.  The frequent deployments of Army personnel in peace operations have led to a multitude of problems, such as higher-than-expected wear and tear on equipment, difficulties with retention of personnel in some units, and (probably the best known problem) the complications in the Army's readiness for potential major theater war (MTW) contingencies as a result of the demands posed by numerous peace operations deployments.[7]  At the most fundamental level, it appears that many of the conflicts the Army will confront in the foreseeable future will be intrastate operations, with traditional MTWs becoming the exception rather than the rule.  Such a prospect raises questions about the nature of the service and the acceptance of peace operations by Army personnel as a core mission for the Army.[8]

Dealing with the consequences of communitarian conflicts is not an optimal way to address the problem of ethnic strife.  A better under-standing and anticipation of such conflicts, which consequently improves the prospects for preemptive remedial action short of using force, is a much better alternative.  In short, preventing strife is almost always a more efficient strategy than dealing with the conse-quences of strife.

Unlike interstate wars, the majority of which end in a negotiated set-tlement, the majority of intrastate conflicts end with the extermina-tion, expulsion, or complete surrender of one side.[9]  And civil wars

[7]Jennifer Morrison Taw, David Persselin, and Maren Leed, *Meeting Peace Operations' Requirements While Maintaining MTW Readiness*, Santa Monica, CA: RAND, MR-921-A, 1998.

[8]Volker C. Franke, "Warriors for Peace: The Next Generation of U.S. Military Leaders," *Armed Forces and Society*, Vol. 24, No. 1, Fall 1997, pp. 33–57; Deborah D. Avant, "Are the Reluctant Warriors Out of Control? Why the U.S. Military Is Averse to Responding to Post–Cold War Low-Level Threats," *Security Studies*, Vol. 6, No. 2, Winter 1996–97, pp. 51–90. Survey research shows mixed attitudes among U.S. Army personnel toward participation in peace operations: see David R. Segal and Ronald B. Tiggle, "Attitudes of Citizen-Soldiers Toward Military Missions in the Post–Cold War Years," *Armed Forces and Society*, Vol. 23, No. 3, Spring 1997, pp. 373–390; Laura L. Miller, "Do Soldiers Hate Peacekeeping?  The Case of Preventive Diplomacy Operations in Macedonia," *Armed Forces and Society*, Vol. 23, No. 3, Spring 1997, pp. 415–449; and David R. Segal, Brian J. Reed, and David E. Rohall, "Constabulary Attitudes of National Guard and Regular Soldiers in the U.S. Army," *Armed Forces and Society*, Vol. 24, No. 4, Summer 1998, pp. 535–548.

[9]In a survey of civil wars in the 20th century, I. William Zartman notes that less than a third have "found their way to negotiation, whereas more than half of modern inter-

with a communitarian or ethnic dimension are especially difficult to negotiate and the most likely to result in protracted strife, often going on for years and sometimes decades.[10] The reason is straight-forward: to end intrastate strife the warring sides must lay down arms and respect an agreement usually in the absence of a legitimate government and under conditions in which the agreement is generally unenforceable.[11] In conditions of communitarian strife, where issues of identity are intertwined in the conflict (since ethnic bonds are psychologically similar to kinship bonds and involve perceptions of identity), it is especially difficult for the two sides to go on coexisting in the same state. Put differently, there are only two main pathways for the regulation of ethnic conflict: (1) eliminating the differences (there are four methods for accomplishing this: genocide, forced transfer of population, partition/secession, and integration/ assimilation); (2) managing the differences (again, four main methods: hegemonic control, arbitration by third party, cantonization/ federalization, and consociationialism/power-sharing).[12] Because trust that would allow for the management of the differences is in short supply once conflict starts, it is no accident that elimination of the differences becomes the preference and that many ethnic and communitarian conflicts end up in prolonged and bloody strife, sometimes mixed in with attempts at genocide and complete elimination of the other side.[13]

---

state wars have done so." I. William Zartman (ed.), *Elusive Peace: Negotiating an End to Civil Wars*, Washington, D.C.: The Brookings Institution, 1995, p. 3.

[10]Parties engaged in an intrastate conflict develop vested interests in the continuation of the conflict, since the strife creates an alternative system of power and profit. David Keen, *The Economic Functions of Violence in Civil Wars*, Adelphi Paper 320, International Institute for Strategic Studies, Oxford and New York: Oxford University Press, 1998. Also see John McGarry and Brendan O'Leary (eds.), *The Politics of Ethnic Conflict: Case-Studies of Protracted Ethnic Conflicts*, London: Routledge, 1993; Elizabeth Crighton and Martha Abele Mac Iver, "The Evolution of Protracted Ethnic Conflict: Group Dominance and Political Underdevelopment in Northern Ireland and Lebanon," *Comparative Politics*, Vol. 23, No. 2, January 1991, pp. 127–142.

[11]Barbara F. Walter, "The Critical Barrier to Civil War Settlement," *International Organization*, Vol. 51, No. 3, Winter 1997, pp. 335–364.

[12]McGarry and O'Leary, *The Politics of Ethnic Conflict*, p. 4.

[13]As many scholars and analysts have noted, ethnic conflict is a type of social competition that "typically involve[s] large amounts of violence and often particularly vicious forms of violence." Samuel Huntington, "Civil Violence and the Process of Development," in *Civil Violence and the International System*, Adelphi Paper 83, London: IISS, 1971, p. 13. In other studies, civil war has been empirically linked to the

Because of the unenforcibility of an internal agreement to end intrastate conflict, third-party intervention is usually required to guarantee the agreement[14] and, even then, the intervening forces easily may become caught up in the continuing struggle between the belligerents.[15] But without an intervention, the simmering intrastate strife may well spawn an international crisis, either in the form of a humanitarian disaster or because a neighboring state becomes drawn into the internal strife and, as a result, creates a regional conflict and the potential for an interstate war.[16]

Thus, for reasons of preventing long-term strife that may escalate to major regional problems, prevention is the preferred course of action. From an economic perspective, the costs of dealing with an ongoing conflict, as well as the opportunity costs and the costs of postconflict reconstruction, are uniformly far greater than the small costs entailed by prevention.[17] And, in the sense of keeping clear of

---

onset of genocides: Matthew Krain, "State-Sponsored Mass Murder: The Onset and Severity of Genocides and Politicides," *Journal of Conflict Resolution*, Vol. 41, No. 3, June 1997, pp. 331–360.

[14]Statistical analysis of all negotiated settlements of intrastate wars in 1945–1997 shows that the most extensively institutionalized settlements (often requiring third-party intervention) are the most likely to prove enduring. Caroline A. Hartzell, "Explaining the Stability of Negotiated Settlements to Intrastate Wars," *Journal of Conflict Resolution*, Vol. 43, No. 1, February 1999, pp. 3–22.

[15]Some have argued that intervention in ethnic conflicts is bound to fail (making the issue of prevention of such conflicts even more important) and see population transfer as the only solution to protracted communitarian strife: Daniel L. Byman, "Divided They Stand: Lessons About Partition from Iraq and Lebanon," *Security Studies*, Vol. 7, No. 1, Autumn 1997, pp. 1–32; Chaim Kaufmann, "Intervention in Ethnic and Ideological Civil Wars: Why One Can Be Done and the Other Can't," *Security Studies*, Vol. 6, No. 1, Autumn 1996, pp. 62–100; Chaim Kaufmann, "Possible and Impossible Solutions to Ethnic Civil Wars," *International Security*, Vol. 20, No. 4, Spring 1996, pp. 136–175; and Chaim Kaufmann, "When All Else Fails: Ethnic Population Transfers and Partitions in the Twentieth Century," *International Security*, Vol. 23, No. 2, Fall 1998, pp. 120–156.

[16]Especially in areas where an ethnic group inhabits both sides of the border, there are strong pressures for an intervention in support of the group, even if the gains are unlikely to be lasting. A. Bikash Roy, "Intervention Across Bisecting Borders," *Journal of Peace Research*, Vol. 34, No. 3, 1997, pp. 303–314. For a broader look at the international implications of communitarian strife, see Michael E. Brown (ed.), *The International Dimensions of Internal Conflict*, Cambridge: Center for Science and International Affairs/MIT Press, 1996.

[17]For an analysis of the costs incurred through lack of prevention and nonintervention, see Nick Killick, "The Cost of Conflict," in Peter Cross (ed.), *Contributing to Preventive Action*, Baden-Baden: Nomos Verlagsgesellschaft, 1998, pp. 97–119; and

military involvement in communitarian conflicts, the United States has a vested interest (quite apart from any moral concern) in preventing the evolution of communitarian and ethnic tensions into outright strife.

However, prevention is easier said than done.[18] Even in cases of a clear drift toward intrastate strife, it may not be possible to marshal the resources required to head off the conflict because, in the absence of a direct threat, it is difficult to expend substantial resources to thwart what some may see as a phantom threat. The general principle is that the earlier the warning, the fewer the resources required and the easier it is to prevent an escalation of tensions to open strife. For example, the prevention of an imminent eruption of violent strife may require the mobilization of substantial diplomatic, economic, and even military resources. But dealing with early signs of a drift toward violence may require only a fraction of what may be needed later. Thus, the role of accurate intelligence and early warning emerges as crucial in the prevention of communitarian and ethnic strife. Accurate early warning cannot take ambiguity and uncertainty out of the process of conflict prevention, and in and of itself it is not sufficient to prevent a conflict, but it is a necessary first step.

The U.S. intelligence community has a mixed record when it comes to anticipating the incidence of communitarian strife. According to some in the intelligence community, they generally "get it right"[19] when it comes to imminent conflict. But the rate of success is not uniform. For example, few predicted the rapid disintegration of Rwanda into genocide. And even when accurate intelligence fore-

---

Reiner Huber, "Deterrence in Bosnia-Herzegovina: A Retrospective Analysis of Military Requirements Before the War," *European Security*, Vol. 3, No. 3, Autumn 1994, pp. 541–551.

[18]There is a large literature and a variety of efforts (among governments, international organizations, and NGOs) to make prevention of conflict more effective. For some of the pitfalls to effective preventive action, see Michael S. Lund, "Preventing Violent Conflicts: Progress and Shortfall," in Cross (ed.), *Contributing to Preventive Action*, pp. 21–63. Also see William J. Dixon, "Third-party Techniques for Preventing Conflict Escalation and Promoting Peaceful Settlement," *International Organization*, Vol. 50, No. 4, Autumn 1996, pp. 653–681.

[19]Robert D. Vickers, Jr., "The State of Warning Today," *Defense Intelligence Journal*, Vol. 7, No. 2, Fall 1998, pp. 9–15.

casts were available, such as regarding the former Yugoslavia, important elements of the driving forces of the conflict were unexpected. In both cases, Army units deployed as part of outside interventions designed to ameliorate some of the consequent suffering. When it comes to accurate strategic warning, that is, anticipating early the likelihood of communitarian strife, the record is also unimpressive. And yet, as mentioned above, for purposes of prevention, it is precisely the early identification of a drift toward violence that is most needed.

This report addresses the issue of early identification of the potential outbreak of communitarian strife. The report is the final product of a project that sought to help Army intelligence analysts and, more broadly, the U.S. intelligence community in monitoring and anticipating more accurately the worldwide potential for intrastate conflict. The research effort sought to improve the Army's anticipatory skills by providing a theoretical model, grounded in the social sciences, of the social processes and dynamics that lead to ethnic and communitarian conflict and state breakdown. Under what conditions are ethnic groups likely to take up violence against the state in order to accomplish their goals? When are they more likely to favor the peaceful pursuit of group aims? Similarly, under what conditions are states likely to resort to violence and repression as opposed to negotiation with the aggrieved group? Put more simply, the project sought to pull together the existing scholarly knowledge about the evolution of communitarian and ethnic conflict into a practical tool for analysts and policymakers. The report is a contribution to the efforts of the intelligence community to come up with better tools to anticipate the incidence of ethnic conflict.[20] In a larger sense, the project aimed to improve the Army's ability to identify and plan for potential conflict contingencies around the world as well as to contribute to interagency planning for such contingencies.

---

[20]In the 1990s, the U.S. intelligence community began to pay much more attention to the issues of intrastate conflict. Classroom hours devoted to operations other than war for intelligence officers have risen sharply since 1990 for all of the services as well as for the DIA (73 percent increase for the Army between 1990 and 1995). Stephen L. Caldwell, "Defense Intelligence Training: Changing to Better Support the Warfighter," *Defense Intelligence Journal*, Vol. 4, No. 2, Fall 1995, pp. 83–100; "Ethnic Conflict— Threat to U.S. National Security?" *Defense Intelligence Journal*, Vol. 1, No. 2, Fall 1992.

## THE MODEL

The theoretical model[21] proposed here differs from most of the intelligence community's existing aids that the authors are aware of and fills what we see as a void. The aids most commonly used take the form of risk and vulnerability assessments. Generally, these are large-scale inductive models that statistically process enormous amounts of data—ranging from health and mortality statistics to the number of individuals under arms—in an effort to develop useful predictors of political violence.[22] Such predictors, if at all available, merely describe correlations between some class of data and political violence. They do not establish a *causal* link between the variables included and the social outcome they seek to explain. Not surprisingly, such inductive models cannot ask questions that bear on the problem of *how* deprivation and discontent lead to strife. Thus they cannot generate a targeted set of information requirements that intelligence agencies can pursue to increase the understanding of causes and the ability to predict with respect to the problem of communitarian violence. Sequential models and especially conjunctural models link a series of events and are more sophisticated tools for early warning of conflict than the widely used correlational models described above.[23] All of these models can play a useful—and increasingly accurate—role in identifying the countries "at risk." But

---

[21]The term "model" is used throughout this report in the conventional sense familiar to the social sciences. It represents a closed system of causal statements providing a theoretical explanation of the phenomenon—communitarian and ethnic violence—under consideration. For an analysis of the concept of a theoretical model, see Max Black, *Models and Metaphors*, Ithaca, NY: Cornell University Press, 1962.

[22]The reports of the State Failure Task Force represent such a tool. See Daniel C. Esty, Jack A. Goldstone, Ted Robert Gurr, Pamel T. Surko, and Alan N. Unger, *Working Papers: State Failure Task Force Report*, November 30, 1995; and Daniel C. Esty, Jack A. Goldstone, Ted Robert Gurr, Barbara Harff, Marc Levy, Geoffrey D. Dabelko, Pamela T. Surko, and Alan N. Unger, *State Failure Task Force Report: Phase II Findings*, July 31, 1998. For references to other models, see Pauline H. Baker and John A. Ausink, "State Collapse and Ethnic Violence: Toward a Predictive Model," *Parameters*, Spring 1996, pp. 19–31. For a survey of some of the methods for early warning of intrastate conflicts, see Pyt Douma, Luc van de Goor, and Klaas van Walraven, "Research Methodologies and Practice: A Comparative Perspective on Methods for Assessing the Outbreak of Conflict and the Implementation in Practice by International Organisations," in Cross (ed.), *Contributing to Preventive Action*, pp. 79–94.

[23]For some problems with each of these tools, see Peter Brecke, "Finding Harbingers of Violent Conflict: Using Pattern Recognition to Anticipate Conflicts," *Conflict Management and Peace Science*, Vol. 16, No. 1, 1998, pp. 31–56.

such models have shortcomings (often a high level of false positives), and they generally fall short when it comes to a detailed consideration of specific cases that might require a closer look at their potential for communitarian strife.

In contrast, the methodology in this report seeks to elucidate how ethnic attachments are channeled into political action that may result in strife. The model provides a conceptual overview of the processes triggering political mobilization along ethnic lines as well as a framework for understanding state responses, both of which interact to produce either political reconciliation leading to peace or political breakdown resulting in violence. The model draws on and incorporates insights from a variety of theories into a more comprehensive structure covering the entire *process* of ethnic mobilization and ethnic challenge to state power. The result is not a "model" in the standard (mechanistic) sense used by intelligence analysts, but rather a method in the tradition of scientific inquiry that aims to explore the basic structure and processes of a given phenomenon. It is, emphatically, not meant to be a mechanistic substitute for the knowledge, reasoning, and judgment of intelligence analysts, but rather a means for helping them order the information already at their command (and identify important information gaps) as they attempt to assess the prospects for communitarian strife.

There are several important considerations to be borne in mind when scrutinizing this model. First, the model is intended primarily to help the intelligence community *order its thinking* about the logic and dynamics of potential ethnic violence, and to systematically organize the *information-collection requirements* relating to the problem. Understanding this intention is critical to appraising the adequacy of the effort.

Second, this model is intended to provide a *general conceptual framework* which speaks to the issue of how ethnic mobilization occurs and how it could lead to violence under certain conditions. It incorporates the insights offered by various theories that focus on separate aspects of the problem (such as, for example, the relative deprivation of the populace or the extent of state capacity) into a more comprehensive structure that encompasses the entire *process* of ethnic mobilization from the roots of conflict all the way to social reconciliation or state breakdown. Having said this, however, it is important to recognize that this framework does not proffer any specific "theory" of ethnic conflict, understood as explaining "why" ethnic conflicts occur in some "ultimate" sense. Rather, the framework offered here is fundamentally an "analog" model in the sense that it aims to identify what, if any, step-level disturbances must occur before the prevailing political system is transformed from one formal operating state into another.

This objective is particularly appropriate for purposes of "indication and warning," where the focus is not so much to divine why ethnic conflicts arise in some "essentialist" way but rather to identify which critical variables are relevant for anticipating its outbreak and what might happen if some of these variables disturb some other variables in the explanatory system. While this task no doubt requires some implicit understanding of the causal drivers of conflict, the emphasis nonetheless is *not* on explicating the various causes and patterns of ethnic conflict per se but instead on developing a "model" that reproduces the structure and relationships within the process of ethnic mobilization in an eidetically adequate way.[24]

Third, this model is focused primarily on helping the intelligence community with the problem of *long-range assessment of possible ethnic strife* rather than with forecasting imminent ethnic violence and state breakdown. While the model speaks to some of these latter issues, it lacks the level of detail required to provide adequate predictions of such violence under a variety of conditions. Incorporating such detail would require more "intensive" sorts of explanations and would be very useful for intelligence analysts tasked with monitoring day-to-day flows of events. However, such models would by definition presume that ethnically driven competition is already under way and, hence, would be less useful from the perspective of trying to assess when and under what conditions ethnic mobilization may in fact come about (just the kind of work that forms the staple of analyses focusing on various long-range futures facing a given state or region). It is also important to recognize that for a variety of theoretical reasons it is probably impossible to develop any *single* model

---

[24]This is precisely the meaning of an "analog" model as opposed to a "theoretical" model. For further detail about the differences between the two structures of scientific explanation, see Max Black, *Models and Metaphors*, Ithaca, NY: Cornell University Press, 1962, p. 223.

which can explain imminent ethnic breakdown. Instead, multiple models would probably be required, and these models would vary based on the outcomes sought to be explained (that is, mass refugee movements, ethnic violence, genocide, and state collapse, for example, would each require separate models) as well as on the relative weights assigned to various remote causes, proximate causes, and event sequences that combine to produce such outcomes. These issues are of lesser concern to an analyst focusing on long-range assessment, who is the primary consumer of this research.

It is also worth mentioning that while the model is intended to provide a general framework for understanding the key structural factors that could lead under certain conditions to ethnic strife, it is *not* designed as a computational device that "automatically" produces predictions of ethnic violence given suitable information. It is in fact—emphatically—not intended to be a mechanistic substitute for knowledge, reasoning, or judgment on the part of regional intelligence analysts but "merely" a means of helping them order the information already at their command (while identifying which information not yet available might be required) as they attempt to assess the prospects for ethnic breakdown in any given region or part of the world.

## ASSUMPTIONS AND CONSIDERATIONS

Readers should keep in mind several considerations and assumptions underlying the theoretical model. Perhaps most important is the whole notion of "ethnic" violence. The plea for disaggregation of the term made by two sociologists is worth stating here:

> The temptation to adopt currently fashionable terms of practice as terms of analysis . . . ought to be resisted. The notion of "ethnic violence" is a case in point—a category of practice, produced and reproduced by social actors such as journalists, politicians, foundation officers, and NGO representatives, that should not be (but often is) taken over uncritically as a category of analysis by social scientists. Despite sage counsel urging disaggregation, . . . too much social scientific work in this domain . . . involves highly aggregated explananda, as if ethnic violence were a homogeneous substance varying only in magnitude. To build a research program around an aggregated notion of ethnic violence is to let public

coding—often highly questionable, as when the Somali and Tadjik-
istani civil wars are coded as ethnic—drive sociological analysis.[25]

We share the dissatisfaction with the general use of the term "ethnic
conflict." But rather than disaggregating, the approach adopted here
is to subsume "ethnic conflict" into the overall process of socio-
political competition. Thus, the terms communitarian conflict or
even intrastate conflict are much preferable to the term "ethnic
conflict." Ethnicity and ethnic attachments play a role to some extent in most intrastate conflicts. But the tendency to identify ethnicity as the cause of the conflict is an unwarranted leap in logic that
good analysis should avoid. Communal differences by themselves
do not provoke conflict.

It is unfortunate that impressionistic and journalistic, rather than
analytical, accounts of the Yugoslav breakup, with their imagery of
brutality and "ancient hatreds" as driving forces, have had so much
influence in public thinking about recent communitarian conflicts.[26]
Such unsophisticated explanations also have had a pernicious
impact on the discourse within the U.S. policy community, with the
"ancient hatreds" explanation even sneaking into President Clinton's
speeches on the topic. It is worthwhile to mention that virtually no
scholar of nationalism and ethnicity accepts the "ancient hatreds" or
the "uncorking of long-suppressed perceptions" explanation of the
post–Cold War communitarian strife. As two prominent political
scientists have noted:

> The most widely discussed explanations of ethnic conflict are, at
> best, incomplete and, at worst, simply wrong. Ethnic conflict is not
> caused directly by intergroup differences, "ancient hatreds" and
> centuries-old feuds, or the stresses of modern life within a global
> economy. Nor were ethnic passions, long bottled up by repressive
> communist regimes, simply uncorked by the end of the Cold War.[27]

---

[25]Rogers Brubaker and David D. Laitin, "Ethnic and Nationalist Violence," *Annual
Review of Sociology*, Vol. 24, 1998, pp. 423–452 (p. 446).

[26]Robert D. Kaplan, *Balkan Ghosts: A Journey through History*, New York: St. Martin's
Press, 1993.

[27]David A. Lake and Donald Rothchild, "Containing Fear: The Origins of Management
of Ethnic Conflict," *International Security*, Vol. 21, No. 2, Fall 1996, pp. 41–75 (p. 41).

Issues of identity have served on occasion as a marker to delineate the opposing sides of communitarian strife. Since some kind of marker is necessary for communitarian competition and strife, ethnicity and ethnic attachments thus have played the role of a necessary component in the strife. But necessary does not mean sufficient. As it is argued here, it takes substantial effort to transform issues of identity onto the platform of political competition. It is the drivers of such transformation that are the crucial factors in leading to communitarian strife, and this report looks at them in detail. Because of its wide usage, the term "ethnic conflict" is used throughout this report. But the reader should keep in mind that the term is used here as shorthand for communitarian intrastate conflict.

Closely connected to the above point, many popular accounts have stressed the eruption of ethnic conflict as primitive, atavistic, and "irrational." But such beliefs are often less explanations than despairing reactions to a difficult phenomenon. If human behavior is irrational, of course it cannot be predicted or even anticipated. If, on the other hand, communitarian strife fits into the general category of social and political competition, its dynamics may indeed be understandable, and so more effective anticipation of ethnic conflict may be possible. In line with the fundamental assumptions of modern social science, the model presented here assumes that human behavior in collectivities is rational and can be understood. Rationality should not be understood as a universally agreed-upon mindset but a recognition that individuals are goal-oriented and adaptive, and that they will attempt to reach their goals by what they see as the easiest and least costly (or most efficient) means. The rationality assumption does not mean that all individuals have the same goals. However, if we understand the goal of an individual, then his actions should be predictable in principle.[28]

The term "ethnicity," because of its fuzziness, is discussed at length at the beginning of Chapter Two. But because of its importance as a central concept in this report, it is worthwhile to summarize here the main idea of "ethnicity." First, ethnicity is not some kind of a given but rather a constructed social phenomenon. As used here, ethnicity

---

[28]For a longer explanation of the rationality idea, as generally used in the social sciences, see Bryan D. Jones, "Bounded Rationality," *Annual Review of Political Science,* Vol. 2, 1999, pp. 297–321.

refers to the idea of shared group affinity and a sense of belonging that is based on a myth of collective ancestry and a notion of distinctiveness. The group in question must be larger than a kinship group, but the myth-engendered sense of belonging to the group stems from constructed bonds that have similarities to kinship. Psychologically, for the individual, the ethnic group is the largest extension of the family. The constructed bonds of ethnicity may stem from any number of distinguishing cultural or physical characteristics, such as common language, religion, or regional differentiation. As a form of constructed identity, ethnicity is more malleable than is often assumed.[29] It is also more ephemeral: it must be continually created and recreated through a multitude of socialization processes. Although ethnic activists often claim that ethnicity is somehow predetermined, it is difficult to subscribe to this claim from a rationalist perspective. However, ethnic attachments are no less real for being socially constructed; indeed, when internalized deeply, ethnic attachments may elicit intense loyalty.

The construction of ethnicity is a by-product of politics, with politics understood as that activity relating to the production of order in social life. Politics forces individuals to discover common resources in their struggle for survival, and the construction of strong bonds based on perceived shared traits amounts to just such a resource, as the bonds then lead to the creation of an "in group" in the ongoing political struggle. Modernization acts as a catalyst for the process of constructing ethnicity, since it forces individuals to operate in a larger social environment than traditional society. In a traditional society, where an individual's "world" is geographically and psychologically limited to a village and its surroundings, bonds based on kinship are sufficient. But when an individual has to deal with impersonal state and market structures and the larger "world" of the state or province, the old bonds no longer suffice. Then ethnicity becomes a useful resource for an individual in his attempt to survive and prosper in a larger social sphere. And in addition to promoting ethnicity, modernization acts as a catalyst to ethnic tensions through homogenization of values and expectations. As socialization pro-

---

[29]Empirical studies have shown repeatedly that group identification is flexible and opportunistic. Robert Grafstein, "Group Identity, Rationality and Electoral Mobilization," *Journal of Theoretical Politics*, Vol. 7, No. 2, 1995, pp. 181–200.

cesses become more uniform, all individuals begin to value the same things, such as wealth, political prominence, and social recognition. Thus, they are brought into conflict with other individuals and ethnic groups over the same bones of contention.

By the same token, ethnicity can be a useful tool for political mobilization. Indeed, the model assumes that ethnic action does not occur spontaneously but rather requires mobilization and direction. This assumption runs against the popular image of a disadvantaged group rebelling spontaneously against state tyranny—a romantic image not borne out in reality. There are many examples of severe group deprivation and repression that do not lead to rebellion, because the group is not mobilized for political action. Without mobilization, ethnically centered perceptions of injustice may exist but do not have larger political significance.

## ORGANIZATION

Chapter Two develops the process model for anticipating ethnic and communitarian strife. First, the chapter elaborates the main approaches to ethnicity—the primordial, the epiphenomenal, and the ascriptive—pervading the literature. Much confusion has resulted from imprecise or different definitions of "ethnicity." The authors make the assumptions explicit and present the framework on which the theoretical model builds. Developing out of the ascriptive tradition, the authors explicate the theoretical model designed to help the intelligence community anticipate the outbreak of ethnically based forms of strife. It explains how the potential for strife should be understood; how the potential for strife is transformed, through mobilization, into a likelihood of strife; and how extant state capacities interact through a process of strategic bargaining with mobilized groups to produce, under certain conditions, varying degrees of strife.

In subsequent chapters the model is applied to four cases. The first two cases are "retrospective." In Chapters Three and Four, the model is applied from the perspective of what an intelligence analyst might have deduced had she used the model to structure analysis of Yugoslavia and South Africa in the late 1980s. Two applications of the model to "prospective" cases follow. Chapters Five and Six apply the model to contemporary Ethiopia and Saudi Arabia so as to analyze the propensity of these states toward major communitarian strife. Finally, Chapter Seven offers some observations on the use of the model, briefly identifies its limitations, and sketches out paths for its further development.

The appendix provides a specific tool for use by analysts to help them order and organize the information-collection requirements into a coherent whole on which to base an assessment. Drawing on the model, it identifies explicitly the questions and indicators that an analyst should consider with respect to the various developmental stages of ethnic tensions and strife.

The model and the questions for analysts (in the appendix to this report) were published originally in 1997.[30] The many reviews of the model in the academic literature and the reactions within the intelligence community have been overwhelmingly positive. A common suggestion has been that the model should be tested. This report documents the results of those tests and provides examples for analysts on how to use the model. We remain convinced of the usefulness of the design and encourage its further development by others.

---

[30]Ashley J. Tellis, Thomas S. Szayna, and James A. Winnefeld, *Anticipating Ethnic Conflict*, Santa Monica, CA: RAND, MR-853-A, 1997.

# Chapter Two

# THE PROCESS MODEL FOR ANTICIPATING ETHNIC CONFLICT

## Ashley J. Tellis, Thomas S. Szayna, and James A. Winnefeld

## APPROACHES TO ETHNICITY

The definition of ethnicity remains one of the most contested issues in social science. Due in part to the definitional problem, the relationship between ethnicity and other classificatory categories such as race, nation, and class remains poorly understood. In fact, the pioneer of modern sociology, Max Weber, even while producing one of the most sophisticated social scientific frameworks that could be applied to analyze ethnicity, concluded that "the notion of 'ethnically' determined social action subsumes phenomena that a rigorous sociological analysis . . . would have to distinguish carefully . . . [and] . . . it is certain that in this process the collective term 'ethnic' would be abandoned, for it is unsuitable for a really rigorous analysis."[1] Considerable scholarly attention to the phenomenon over the past three decades seemingly has been intent on proving Weber wrong. The more recent analyses of ethnicity have taken a variety of approaches and produced some sophisticated new insights. But when these disparate approaches are analyzed systematically, they usually fall into one of three main "ideal types" relating to ethnicity as a social phenomenon, each with a distinctive perspective on what ethnicity is and how it relates to social conflict.

These three approaches are discussed below. First, each approach's principal insights are systematically described, along with its particu-

---

[1]Max Weber, *Economy and Society*, Vol. 1, Guenter Roth and Claus Wittich (eds.), Berkeley: University of California Press, 1968, pp. 394–395.

lar explanation of the linkages between ethnicity and group conflict. The strengths and weaknesses of each approach are then analyzed with respect to group solidarity and collective action.  "Group solidarity" refers to the existence of deep bonds between members of a particular social grouping.  "Collective action" refers to the basic problem facing any group action, namely, the fact that contributing to collective action may not be rational for an individual and, as such, may result in the failure of group efforts.

## The Primordialist Approach

The first approach to ethnicity is commonly termed the "primordialist" approach in that it centers on the assertion that certain primitive (or basic) sociological groupings exist in a society. Such primitive groupings exist *a priori*, meaning that they are natural units that derive their cohesion from some inherent biological, cultural, or racial traits which then become instruments of social differentiation.  The primordialist school asserts that human societies are in effect conglomerations of "tribes."[2]  The regulating principles that define the distinctions between "tribes" may vary, but what is crucial is that they determine both the boundaries and the meaning of tribal membership in such a way that the "in-group" and "out-group" can always be clearly demarcated.  Such *a priori* groupings then constitute the primitive units in society:  they define for their members critical existential distinctions centered on the dichotomy of "us and them"[3] and they perform the crucial task of forming an individual's personal identity through a process of "collective definition."[4]  This process, defining the way that "racial groups come to see each other and themselves," relies on a constant redefinition and reinterpretation of historical events and social experiences vis-à-vis other such groups, and it eventually results in the "aligning and realigning of relations and the development and reformulation of prospective

---

[2]Harold Isaacs, *Idols of the Tribe:  Group Identity and Political Change*, New York: Harper Row, 1975, pp. 39–45.

[3]Frederik Barth, "Ethnic Groups and Boundaries," in Frederik Barth (ed.), *Process and Form in Social Life:  Selected Essays*, London:  Routledge & Kegan Paul, 1981, pp. 198–227.

[4]H. Blumer and T. Duster, "Theories of Race and Social Action," in *Sociological Theories:  Race and Colonialism*, Paris: UNESCO, 1980, p. 220.

lines of action toward one another."[5] Thus, in the primordialist view, ethnic groups function as insular universes. Their membership is defined by accident of birth, and once constituted, they perpetuate their distinctiveness by a continuing process of socialization that accentuates their perceptions of uniqueness and their sense of separateness from other, similar, social formations.

**Primordialists and conflict.** It is important to recognize that in the primordialist approach, ethnicity is *not* a problematic social category. It is, in fact, almost self-evident, and as a result its theory of conflict is also relatively simple and easy to discern. Although there are as many theories of conflict among the primordialists as there are primordialists themselves, the general logic of conflict generation runs somewhat along the following lines: Ethnic groups are located in pluralist societies that contain several other similar competing social formations. Although relations *within* the ethnic group may be either personal or impersonal in nature, social relations *between* ethnic groups in large societies are invariably impersonal and usually take place through market structures or the political process. These institutions, concerned as they are with the production and distribution of wealth and power, invariably create both winners and losers. To the degree that the winners and losers are congregated disproportionately within some ethnic group, opportunities arise for interaggregational conflict leading to violence. Even if a given ethnic group does not host a disproportionately large number of *objectively* disadvantaged individuals, interaggregational conflict could still occur if, through the process of collective definition, the group internalizes a "myth" of deprivation, thereby channeling individual resentments toward other groups rather than diffusing them within itself. Conflicts as a result of competition over resources certainly could occur even within an ethnic group, but the primordialist assumption—that ethnic groups are characterized by strong forms of organic solidarity flowing from self-evident ties of biology, culture, or race—implies that such competition either would not be significant or would not result in large-scale violence directed at one's ethnic cohorts. Such significant, large-scale violence would almost by definition be directed primarily at other ethnic groups, justifiably or

---

[5]Ibid., p. 222.

not, for reasons connected with the larger competitive—tribal—constitution of society.

**Primordialists: strengths and weaknesses.** The primordialist approach to ethnicity has one important strength, but it suffers from many weaknesses. Its singular strength stems from its focus on factors that easily explain social solidarity. There is little doubt that in many societies, superficial human similarities like pigmentation, shape of hair, and other such physical characteristics often serve as elementary justification for simple forms of social solidarity. This solidarity is dependent on a popular belief in a common ancestry based on the notion of "they look like us" and may in fact be reinforced by a common language, common history, and common enemies. Such variables often serve to define a group's identity, but in the primordialist account, they suffice to explain both the nature of group solidarity and how collective-action problems are resolved. While primordialist explanations of the former are easier to swallow (especially when group solidarity takes on superficial manifestations and does not involve either high or asymmetrically distributed individual costs), they are harder to accept with respect to resolving the collective-action problem, except perhapsofeties, or when the "ethnic" groups concerned are extremely small or are in fact simply a form of kinship grouping (such as clans).

In all other cases, where a modicum of egoist motivation is combined with the presence of instrumental rationality, the collective-action problems inherent in any coordinated group action become more difficult to resolve and cannot be waved away by the simple unproblematic assertion of organic solidarity. This is all the more true because of a glaring empirical problem that contradicts primordialist beliefs, namely, that group identities historically have never been fixed, they are constantly changing, and new identities arise all the time. Moreover, even existing "ethnic" groups contain individuals of varying degrees of common ancestry, they change in composition over time (as in, for example, the creation of the "Anglo-Saxons"), and most important, all ethnic groups have to confront the problem of in-group struggles for power that affect the kind of social solidarity they can amass for purposes of effecting successful collective action. All these factors taken together suggest that the nature of ethnic solidarity is itself highly problematic and cannot be produced as effortlessly on the basis of merely superficial human characteristics, myths of common ancestry, or even a shared history, as the primordialist account tends to suggest.

## The Epiphenomenalist Approach

In sharp contrast to the primordialist account, which provides an "essentialist" description of the meaning of ethnicity, the second approach, which might be termed the "epiphenomenalist" perspective, denies that "ethnicity" as a *social* phenomenon has any inherent biological basis whatsoever. This approach to ethnicity, evident especially in the Marxist tradition, does not by any means deny the raw existence of physical or social differences ultimately derived from biology and perhaps finally manifested in some specific cultural forms. However, it denies the claim that such biological or cultural formations have an *independent effect, unmediated by class formations or institutional relationships, on politics.*[6] In fact, the epiphenomenalist perspective emphatically asserts that it is the class structures and institutionalized patterns of power in society that are fundamental to explaining political events rather than any biologically or culturally based social formations like "ethnicity." To the degree that "ethnicity" in the primordialist sense plays a role, it functions merely as a "mask" that obscures the identity of some class formations struggling for political or economic power. Ethnicity per se is, therefore, merely an incidental appearance (i.e., "epiphenomenal"): it is not the true, generative cause of any social phenomenon, even though it often may appear to be.[7]

To the degree that ethnicity appears at all, scholars who accept the epiphenomenalist approach treat it either as a strategy for mobilization on the part of class elites forced to latch on to such means of group identity by pressures of necessity[8] or, especially in Marxist sociology, as a transient form of false consciousness that will be

---

[6]For an excellent survey of Marxist perspectives on race and ethnicity, see John Solomos, "Varieties of Marxist Conceptions of 'Race,' Class and State: A Critical Analysis," in J. Rex and D. Mason (eds.), *Theories of Race and Ethnic Relations,* Cambridge: Cambridge University Press, 1986, pp. 84–109.

[7]R. Miles, "Marxism Versus the Sociology of Race Relations?" *Ethnic and Racial Studies,* Vol. 7, 1984, pp. 217–237.

[8]See, for example, J. Rothschild, *Ethnopolitics: A Conceptual Framework,* New York: Columbia University Press, 1981.

superseded in due time by true class consciousness.[9]  In any event, ethnicity in the primordial sense is denied altogether the status of an "efficient cause." It matters primarily as a "label" that identifies different groups placed in relations of cooperation, symbiosis, and conflict based on their location amid the relations of production in a given society.

**Epiphenomenalists and conflict.** The epiphenomenalist approach to "ethnicity" witnessed in Marxist thought was conditioned in part by the fact that Marxism developed within the more or less culturally homogenous capitalist societies of Western Europe, where class relations rather than ethnic formations become a natural focus for theory.  However, when Marxist ideas spread to more ethnically heterogeneous societies beyond Europe, the issue of ethnicity as a factor in social action had to be addressed anew.  It was only then that theorists provided the definitive reading that integrated ethnicity, as understood by the primordialists, within the larger Marxist theme of generalized class struggle.  Such integration in effect denied that ethnicity could ever be an independent element disturbing the normal, conflictual dynamics in capitalist society, but it did point to three critical features as part of its theory of conflict that any model seeking to assess the prospects for ethnic violence must confront. The Marxist theory of social conflict is too comprehensive and too well known to merit being summarized here.  However, three insights with respect to ethnicity are worth reiterating.

First, any scrutiny of ethnicity with respect to political mobilization would do well to examine the prevailing means and relations of production existing in any society.  Such an examination would identify differential, and perhaps exploitative, patterns of distribution of wealth and power and thus suggest potential "class" groupings that might manifest themselves under certain conditions through "ethnic" terms.

Second, the role of the state in "reproducing" ethnicity is crucial.[10] This implies that the state apparatus may simply preserve what

---

[9]See, for example, R. Miles, "Racism, Marxism and British Politics," *Economy and Society,* Vol. 17, 1988, pp. 428–460, and R. Miles, *Racism,* London:  Routledge, 1989.

[10]See the discussion in Michael Omi and Howard Winant, *Racial Formation in the United States,* New York:  Routledge & Kegan Paul, 1986.

appears to be certain "ethnically" structured patterns of exploitation. Such patterns may be obvious and may even be defined "ethnically" (as in the case of *apartheid* in South Africa), but they are more accurately part of a larger effort at maintaining the economic power of a certain class rather than merely "ideological" conflicts between certain biologically grouped social formations. Under certain conditions, the state may create "ethnicity" as a vehicle of political mobilization. The state may even embark on such efforts in order to assert its own autonomy against certain dominant classes or with an eye to bolstering its prospects for survival against incipiently rising challengers.

Third, "ethnic" violence is invariably (or, at least, often enough) a product of class antagonisms or of class-state antagonism. This implies that ethnic violence is often more than it appears to be. Very rarely, is it—as journalistic descriptions are apt to portray—an accidental or irrational outcome caused either by passions getting out of hand, or the "madness of crowds," or differences in outlooks rooted in religion, *weltanshauung*, or other cultural differences. While such ingredients may constitute the proximate cause of violence in any given situation, the roots of the violence lie in fissures deep beneath the surface—fissures that more often than not are strongly connected with serious and consequential contentions about the distribution of wealth and resources.

**Epiphenomenalists: strengths and weaknesses.** The "epiphenomenal" approach to ethnicity has many strengths, but one important weakness. Its principal strength lies in the fact that it draws attention to politics as a struggle for wealth and profit. Consequently, it provides a fairly coherent account as to why certain empirical groupings can coalesce politically and how they can act collectively with respect to the ongoing competition for resources. Instead of *a priori* claims of biological or cultural affinity as explanations for such solidarity, the epiphenomenalist perspective accounts for it as a product of individual self-interest combined with differential access to the means of production. The epiphenomenalist view thus paves the way for viewing "ethnicity" not as a permanent characteristic of individuals, but as an incidental coloration that may or may not acquire salience depending on the circumstances confronting the process of mobilization. If "ethnicity" becomes salient, however, it can be understood as part of a larger structural competition that engages both interclass and class-state relations. By focusing on these critical arenas of social action, the epiphenomenal view reminds the analyst about the constant struggle for resources in any society and, accordingly, provides a means of contextualizing "ethnic" struggles within the larger realm of social competition.

The epiphenomenal perspective represents a useful corrective to the "essentialist" approach favored by the primordialists, but it has one important weakness, namely, the reliance on a single explanatory variable, such as class in the Marxist approach. This focus on a single variable precludes it from analyzing ethnicity as a phenomenon in its own right. While the importance of ethnicity as an independent causal factor may in fact be small (and Marxism has done more than perhaps any other system of analysis to establish this conclusion), the extent of its contribution and the nature of its appearance in any given case can only be discerned empirically. As such, it still merits independent examination, if for no other reason than the multitude of sins attributed to it. Marxism, however, has few tools to conduct such an examination, since it simply subsumes "ethnicity" under a wider set of social relations or merely treats it as a superstructural phenomenon hiding the "true" base.[11]

## The Ascriptive Approach

The third approach to ethnicity overcomes the limitations of both the primordial and the Marxist perspectives while retaining the best elements of both traditions. This approach, which can be dubbed "ascriptive" and derives from the work of Max Weber and scholars who have followed him, is distinguished by a view of ethnicity that is best described as *real, but constructed*. Weber's discontent with the "multiple social origins and theoretical ambiguities"[12] of ethnicity as a concept have been alluded to earlier, but despite his arguments about the "disutility of the notion of the 'ethnic group,'"[13] he defined the latter carefully enough to make productive analysis possible. Ethnic groups, according to Weber, are

---

[11]Solomos, op. cit.

[12]Weber, *Economy and Society,* p. 387.

[13]Ibid., p. 393.

those human groups that entertain a subjective belief in their common descent because of similarities of physical type or of customs or both, or because of memories of colonization and migration. This belief must be important for the propagation of group formation; conversely, it does not matter whether or not an objective blood relationship exists.[14]

The key distinguishing mark of ethnicity in this reading is the notion of "subjective belief," that is, the animating conviction held by members of a group that they do enjoy a common ancestry and hence are bound together by some ineffable tie that in truth may be wholly fictitious. Thanks to this belief, "ethnic membership differs from the kinship group precisely by being a *presumed identity* . . . ."[15] This insistence on presumed identity as the structuring principle of ethnic unity does *not* imply, however, any particular consequences for social and political action. In fact, Weber insistently argues that "ethnic membership does *not* constitute a group; it only facilitates group formation of any kind, particularly in the political sphere."[16]

Weber's general approach thus consists of the affirmation that ethnic groupings can exist as a result of certain mythical intersubjective beliefs held by a collectivity, but their mere existence does not have any *necessary* consequences for social action. To the degree that the latter phenomenon needs to be explained, the explanations must be found elsewhere. Through such a "constructivist" approach Weber manages to solve the vexing problem afflicting both the primordialist and epiphenomenalist traditions, in that he suggests how "ethnicity" can be treated as a causative variable without allowing it to dominate the explanatory space. As described earlier, the primordialist asks for too much, while the epiphenomenalist gives away too little. The former treats ethnicity as a self-evident fact that automatically explains all manner of collective action; the latter, in contrast, treats ethnicity as merely an incidental appearance that can be readily and completely discarded in favor of more consequential, though hidden, social forces.

---

[14]Ibid., p. 389.

[15]Ibid., p. 389.

[16]Ibid., p. 389.

Weber carves an artful middle ground. He allows for the possibility that ethnicity can be created ("ascribed"), even if only through the collective imagination of apparently similar individuals. In so doing, he could be said to accept an insight that has roots in the primordial-ist tradition. But he does not make the fact of ethnic groups arising determinative for any social acts. Thus he allows for the possibility that precisely those kinds of forces identified by the epiphenomenal-ists might be among the real drivers beneath what are otherwise taken to be "ethnic" phenomena. In fact, he broadens the epiphe-nomenalist insight still further. Instead of restricting himself merely to the class dynamics seen to operate in capitalist societies (which Marxists would assert represent the real movers beneath "ethnic" action), Weber would suggest that the compulsions of politics—understood as the struggle for power writ large—explain the origins and persistence of groups and cause them to discover a variety of solidarity-producing myths like ethnicity. Thus, he notes that "it is primarily the political community, no matter how artificially orga-nized, that inspires the belief in common ethnicity. This belief tends to persist even after the disintegration of the political community, unless drastic differences in the custom, physical type, or, above all, language exist among its members."[17]

**Ascriptivists and conflict.**  The above passage makes clear the ascriptive approach to ethnicity: Politics creates ethnicity in that it forces individuals to discover common resources in their struggles for survival.[18] The fundamental role of politics implies that ethnicity as a phenomenon becomes real only because of the subjective constructions of individuals under certain circumstances and not because it exists *a priori* as some intrinsically permanent solidarity binding a set of individuals across time and space. Such a perspec-tive, then, provokes a set of questions that demand further investiga-tion by any model seeking to anticipate the incidence of ethnic strife. First, what factors precipitate the generation of these intersubjective beliefs relating to ethnicity? Second, how do these intersubjective

---

[17]Ibid., p. 389.

[18]In some sense, the discovery of such associations is a "natural" process because "if rationally regulated action is not widespread, almost any association, even the most rational one, creates an overarching communal consciousness; this takes the form of a brotherhood on the basis of the belief in common ethnicity." Weber, *Economy and Society*, p. 389.

beliefs, once generated, come to precipitate group mobilization and possibly collective action?  And, third, how do state actions—which, as the Marxists correctly point out, are critical to reproducing ethnicity—contribute to group definition, mobilization, and possibly violent social action?  Weber's sociology and the scholars working in his tradition provide pointers that direct the search for answers to these questions.  Three key insights stand out.

First, Weber draws attention to the centrality of power and domination as the fundamental features of politics, the latter understood as that activity relating to the production of order in social life.  Domination, in fact, "constitutes a special case of power"[19] and in a general sense describes all the mechanisms concerned with "imposing one's own will upon the behavior of other persons."[20]  When viewed in such terms, domination becomes "one of the most important elements of social action" and although "not every form of social action reveals a structure of dominancy," it plays a considerable role in all varieties of social action, including the economic, "even where it is not obvious at first sight."[21]  The key insight residing here, accordingly, consists of the claim that politics is central to the regulation of social life.  Politics subsumes economics and, as such, the quest for domination that drives all political life appears in the economic realm as well: "Economic power . . . is a frequent, often purposefully willed, consequence of domination as well as one of its most important instruments."[22]  Because domination is so central to political, economic, and even social structures, any theoretical model that seeks to understand the generation of ethnicity *must* focus on how individuals attempt to structure patterns of domination in political communities, and such structuring of domination must be investigated both with respect to the in-group and the political community at large.[23]

---

[19]Weber, *Economy and Society,* Vol. 2, p. 941.

[20]Ibid., p. 942.

[21]Ibid., p. 941.

[22]Ibid., p. 942.

[23]For a pioneering theoretical analysis of how such patterns of domination are manifested at multiple levels, see Frank Parkin, *Marxism and Class Theory: A Bourgeois Critique,* New York: Columbia University Press, 1979.

Second, since the struggle for domination lies at the heart of all community life, it brings in its trail efforts to monopolize power. These efforts occur both within groups and in society at large in an interactive and often mutually reinforcing fashion. Weber described these dynamics by the phrase "monopolistic closure,"[24] which refers to all efforts made to prevent "others" from sharing in the political, economic, and social bounties enjoyed by a few. These efforts can occur in different ways, depending on historical and social circumstances. They may begin as small confederate efforts mounted by a few individuals to capture power, wealth, or social status. Such efforts would be designed to keep "others" out merely to maximize the distributed share of acquired gains. If these strategies are successful and in fact come to be institutionalized in ethnic terms polity-wide, they would represent forms of "outside pressures" which in turn could impel disadvantaged individuals to search for other ideational devices to use to generate responding forms of resistance. The way these patterns of closure *originate* is less interesting than the fact that they do exist pervasively in all societies: dominant groups use them to keep others out; successful but still relatively disadvantaged groups use them to prevent a dilution of their gains; and utterly disadvantaged individuals tend to use them to recover a modicum of solidarity, which makes resistance and the pursuit of revanchist strategies possible.[25]

Third, the criteria for justifying individual or group efforts at monopolizing power do not have to be objectively defensible. What is important to recognize when exclusion is attempted by any groups in ethnic terms is that "*any* cultural trait, no matter how superficial, can serve as a starting point for the familiar tendency to monopolistic closure."[26] Often, it does not matter which characteristic is chosen in the individual case: whatever suggests itself most easily is seized upon. Consequently, any theoretical model seeking to understand the generation of ethnicity must go beyond proximate ideational labels to scrutinize the underlying structures of domination and

---

[24]Weber, *Economy and Society*, p. 388.

[25]This notion of "dual closure" is examined in some detail in Parkin, op. cit., pp. 89–116.

[26]Weber, op. cit., p. 388.

deprivation in the political, economic, and social realms that make recourse to such labels relevant.

**Ascriptivists: strengths and weaknesses.** The ascriptive approach is attractive because of its multiple strengths. Its primary strength is that it is an open-ended approach: it provides a means of viewing ethnicity that allows for the integration of "primordialist" and "epiphenomenalist" insights, but it does not force the analysis into any single predetermined hypothesis. It identifies core issues that must be engaged—like the instrumental nature of ethnic phenomena, the pervasiveness of domination, and the ubiquity of overt and covert forms of social closure—but it leaves the analyst free to incorporate these factors into any causal hypothesis of his own choosing. Thus, it opens the door for the incorporation of insights deriving from resource competition, rational choice, and symbolic interactionist approaches to ethnicity.[27] Further, it assumes the premise of methodological individualism. It makes rational individuals the theoretical primates for purposes of social explanation, even as it can accommodate macroscopic entities like "ethnic groups" for analytical purposes. Being centered on individuals with specifically attributed preferences and capabilities, it avoids the problems of reification that arise from the use of ungenerated macroscopic wholes. Because of its flexibility, the approach lacks the easily identifiable flaws of the other two approaches.

## Rationale for Using the Ascriptive Approach

For all the reasons examined above, the ascriptive approach to ethnicity is the most defensible approach to developing a model for anticipating ethnic violence. This approach may in fact be vehemently repudiated by ethnic activists, since such individuals often have a deep stake in primordialist conceptions of ethnicity. Similarly, those with a stake in doctrinaire ideological explanations of social relations may also deny the validity of an ascriptive approach. But most social theorists today would admit that an ascriptive approach incorporating both Marxist and Weberian insights is the

---

[27]Several of these approaches with respect to ethnicity are detailed in Rex and Mason, op. cit.

most fruitful avenue to understanding the larger problem of exclusion and domination in society.[28]

In light of the above discussion of various competing approaches, it is worthwhile to restate the concept of ethnicity. The concept entails three crucial components: distinguishing characteristics (any and/or all of the following: phenotype, faith, language, origin, or population concentration in a given region), a sense of group solidarity, and contact with another group so as to establish a point of reference and the idea of "otherness." Ethnicity is defined as the idea of shared group affinity, and belonging is based on the *myth* of common ancestry and a notion of distinctiveness. The group in question must be larger than a kinship group, but the sense of belonging—based on myth—stems from created bonds that have close similarities to kinship. The basis for these created bonds may stem from any number of distinguishing characteristics. These are, at any rate, incidental and case-specific, though they may lead, under certain situations, to deep personal attachments.[29]

## INTRODUCING THE MODEL

In this section we describe our conceptual model for anticipating ethnic violence. The model focuses on how the grievances stemming from existing patterns of domination and deprivation in any society could be translated into imperatives for group mobilization, which in turn interact with various state actions to produce a variety of political outcomes ranging from political reconciliation to state breakdown. This approach attempts to understand the dynamics of group definition, ethnic mobilization, strategic bargaining, and political action as part of a single continuous process.

Taking its cues from the ascriptive tradition, the model described below approaches ethnicity primarily as a "marker," that is, as a real but constructed instrument for defining group identity as a prelude

---

[28]An excellent example of this claim may be found in V. Burris, "The Neo-Marxist Synthesis of Marx and Weber on Class," in N. Wiley (ed.), *The Marx-Weber Debate*, Newbury Park: Sage, 1987, pp. 67–90.

[29]This usage of the term "ethnicity" is based on Donald L. Horowitz, *Ethnic Groups in Conflict*, Los Angeles: University of California Press, 1985, who derives his conception from the larger ascriptive tradition associated with Weber.

to collective mobilization and social action. Thus, the approach seeks to accommodate insights from both the primordial and epiphenomenal schools: It accepts that ethnicity can be used to identify certain social formations, and that ethnicity in this sense can derive from *any* perceived commonalities such as race, language, religion, geographic origins, or culture, in addition to more direct affinities derived from kinship. However, it presumes that such ethnic markers arise principally against a backdrop of ongoing social struggles, which may have conspicuous economic components but are not necessarily restricted to them.

The resulting model, graphically illustrated in Figure 1, focuses on the dynamics of mobilization for the purposes of highlighting the three basic issues critical to the outbreak of ethnically based violence. First, what is the structural potential for strife? Second, what are the requirements for potential strife to be transformed into likely strife? Third, how does likely strife degenerate into actual strife? In

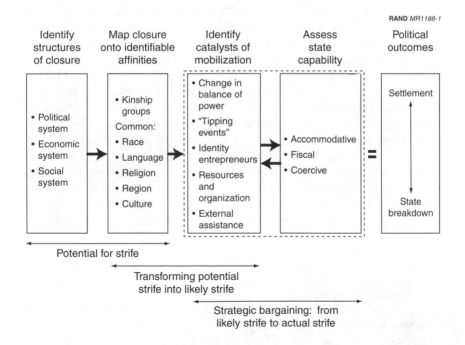

Figure 1—Anticipating Ethnic Conflict

attempting to address these three issues, the model identifies three bilateral interactions. Each successive interaction builds on the foregoing one in an effort to typify the generative process describing the dynamics of ethnic violence.

## STAGE I: INVESTIGATING THE POTENTIAL FOR STRIFE

### Closure

The first issue the model seeks to address is the structural potential for strife. Consistent with the ascriptive tradition, it seeks to determine this potential for strife through an investigation of the patterns of closure existing in a given society. The Weberian concept of closure, which is central here, refers to the "process of subordination whereby one group monopolizes advantages by closing off opportunities to another group."[30] Closure theorists distinguish between two main reciprocal modes of closure: exclusionary and usurpationary. Exclusionary closure involves the exercise of power in a downward direction, when one group attempts to secure its advantages by closing off the opportunities available to another group beneath it by treating the latter as inferior or ineligible. Usurpationary closure involves the exercise of power in an upward direction, as for example when subordinate groups attempt to erode the advantages of higher groups.[31] Both modes represent means for mobilizing power to enhance or defend a group's share of rewards or resources. Because the struggle for power is constant in any society, the justification for exclusion as embodied in various forms of closure is always open to challenge, and the public arena constantly witnesses efforts to make the boundaries between dominator and dominated either obscure or transparent, legitimate or illegitimate.[32] And since the state ultimately rationalizes the legal terms of closure and domination, it becomes the central arena where the processes of exclusion and usurpation unfold.

---

[30]Raymond Murphy, *Social Closure: The Theory of Monopolization and Exclusion*, Oxford: Clarendon Press, 1988, p. 8.

[31]Ibid., pp. 8–10; The concepts of "exclusionary" and "usurpationary" closure were originally developed by Parkin, op. cit.

[32]Murphy, op. cit., p. 48.

**The three realms of closure.** The three arenas of critical importance for the purposes of examining closure are the political, the economic, and the social. The political arena relates to all matters of governance, administrative control, and command over the means of coercion. The economic arena relates to all matters connected with the production of wealth and the distribution of resources. The social arena relates to all matters connected with the effective claims on social esteem, including the distribution of status and social privileges.

Each of these arenas demarcates a particular dimension of the public order and each is related in multiple ways to the others, but constrictions in the political realm are perhaps the most important form of closure for purposes of establishing the potential for strife. The political realm has such primacy because the issue of who rules and how directly affects all kinds of outcomes within the polity.[33]

To begin with, political authority literally holds the power of life and death over its citizenry as a result of its monopolistic claim over the means of coercion and the legitimacy normally accorded to its right to employ force. This power is usually exercised through myriad tools of domestic law enforcement, direct political action against individuals and groups, and through organized state efforts at staving off political disorder. Consequently, the nature of political authority and state structure affects the primary interests of individuals—survival—in a very direct, and often unmediated, way. The political realm also has primacy because it embodies the "rules of the game." It represents the differential and regulated access to power afforded to different groups as well as control over the ability to amend these rules of access.

Finally, the political realm enjoys primacy because it can control—through force if necessary—both rules and outcomes in the economic and social realms. To be sure, economic power often provides the buttress for political power, but it is usually indirect in its effects and invisible as far as its immediacy is concerned. At any rate, holders of economic power cannot survive direct competition with political authority, except by recourse to political means. For this reason,

---

[33]See the discussion in James Buchanan and Gordon Tullock, *The Calculus of Consent,* Ann Arbor: University of Michigan Press, 1962.

economic power holders generally seek to operate in arenas relatively autonomous from state action or attempt to ally with holders of political power for survival and domination.

The social realm is perhaps the least salient of the three arenas, except perhaps in traditional societies. Because status monopolies usually derive from vocational origins, hereditary charisma, or hierocratic claims, the social realm in nontraditional societies usually acquires importance only when the status groups within it are linked to larger loci of economic or political power. Indeed, the derivative nature of the social realm is often evident from the fact that politically or economically advantaged groups tend to have a high social status.

Despite the centrality of the political realm, assessing the potential for strife requires identifying patterns of asymmetrical relative power and domination in all three realms. This is done not only to develop a particular profile of a given country or region, but also to identify in advance which sets of individuals might coalesce, form alliances of convenience, or become pointed antagonists as the process of ethnic mobilization evolves over time. Which of these behaviors will actually occur cannot be predicted *a priori*, in the absence of additional information about the preferences and capabilities of each of these groups, but the very fact of identifying the relevant actors and their location vis-à-vis each other already provides some means of anticipating how different intergroup interactions may evolve over time.

## Identifying Closure

Keeping in mind that the principal objective of the first stage of the model is to identify the prevailing structures of closure in the political, economic, and social realms, there are two important variables in each of these arenas: (1) the existing patterns of distribution with respect to power, wealth, and status respectively, and (2) the relative ease with which individuals can secure access to power, wealth, and status through peaceful means. Each of these variables is important in different ways. The first variable depicts a static picture of existing pattern of dominance and deprivation, whereas the second variable speaks to a more dynamic issue, namely the possibilities for peaceful change.

**The political realm.** At the level of the political system, scrutinizing the existing distribution of power requires assessing the composition of the principal political authorities in the country with respect to their subnational categories like race, language, religion, region, or culture. Such an assessment requires sensitivity to the nature of the political regime and to the question of unitary (one-branch) or divided (multibranch) government. If the government is unitary, attention should be focused primarily on the executive and bureaucratic arms of the state. This requires scrutinizing the composition of the executive decisionmaking elite (whether civilian or military) as well as the upper echelons of the bureaucracy. The former consists mainly of ministerial-equivalent positions and above, whereas the latter relates to director-level positions and above in the civilian bureaucracy and to colonel-equivalents and their superiors in the military and internal security-police bureaucracies.

To the degree that state structure is characterized by divided rather than unitary government, the width of relevant focus groups would have to be expanded. The composition of legislative and judicial bodies might also have to be scrutinized in the context of understanding both the nature of, and the constraints imposed on, the exercise of power in such divided governments. A special circumstance arises in federal states. Because of the devolution of central power implied by a federal setup, in such cases an examination of the type suggested above is warranted at both the state and regional (provincial) levels.

Assembling raw compositional data on how the institutions singled out above are manned in any form of government provides a valuable sociological profile of a state (or province). But such information by itself does not confirm the extent of political closure within the polity. To derive such an assessment, it is important to relate the compositional data about the leadership population to its relative size within the population as a whole. In other words, the actual proportion of political elites measured by subnational categories like race, language, religion, region, or culture should be compared with the size of such demographic categories within the population as a whole. Such a comparison of actual representation vis-à-vis some notionally appropriate number derived from proportional representation may provide some clues about the level of political closure.

An understanding of the *possible* extent of political closure should be complemented by an assessment of whether the state in question pursues a formal agenda of closure aimed at restricting the participation of some groups in political life. This is important because institutional designs and structures have a fundamental bearing on whether patterns of closure derived from numerical comparisons are in fact significant, spurious, or merely accidental. Accordingly, there is a need for additional information about the second variable identified earlier, namely, the relative ease with which individuals can secure access to power through peaceful means. This implies examining a number of factors: whether there are formal or informal restrictions on political participation; whether there is universal access to a system for peacefully redressing grievances; whether ruling institutions and groups subject themselves to periodic, peaceful tests of legitimacy; and whether there is a viable civil society. Information about such dynamic elements of access to power is crucial if one is to judge the nature and extent of political closure. Often such information may not be quantifiable, although there are several examples of attempts to quantify the degree of responsiveness within a political system.[34] Nonetheless, such information is necessary and must be related to other numeric data on representation if a better and clearer understanding of the patterns of political closure within a given society is to be derived. The patterns derived from the first examination might serve as a tip-off for an examination of the formal rules of closure and exclusion that may exist in the society.

**The economic realm.** A similar scrutiny is necessary in the economic realm. Here too, the focus rests on identifying first the general patterns of how wealth is distributed as well as the relative institutional freedom of access to resources existing within the polity. The overall pattern of distributed wealth and income provides an important clue with respect to the issue of economic closure.[35] The objective again is first to assemble data that identify how income and property are distributed at the elite and general population levels, with the data

---

[34]The annual surveys produced by such organizations as Freedom House and Amnesty International are examples of such efforts.

[35]See the discussion in Manus I. Mildarsky, "Rulers and Ruled: Patterned Inequality and the Onset of Mass Political Violence," *American Political Science Review*, Vol. 82, 1988, pp. 491–509.

categorized by reference to certain substate categories like race, language, religion, region, or culture within a given country. The data then must be related proportionally to the relevant demographic categories in the population as a whole (just as in the political data, misrepresentation of demographic categories is likely in many societies). The rationale for advocating analysis at the economic realm at both the elite and the general population levels is to cross-check for the accuracy of the patterns.

The information then must be complemented by perhaps more important qualitative information about institutional structures and about individuals' relative access to economic resources. For example, are there any formal or informal restrictions on investment, trade, business, and employment directed at particular groups categorized by the substatal categories described earlier? Such restrictions, coupled with quantitative data on the distribution of wealth and income, would provide powerful evidence of serious economic closure with consequent implications for the potential for strife.

If the state's economy is not predominantly market oriented but has command features in varying degrees, the second issue of relative freedom of access to resources intersects more clearly with political mechanisms of closure and becomes all the more important. A command economy concentrates wealth in the hands of state managers who can embark on exclusionary distribution schemes in a much more explicit and organized fashion than market mechanisms can. If the principles of exclusion in such arrangements are defined either explicitly or implicitly by certain substatal categories like race, language, religion, region, or culture, the problems of economic closure may be much more intractable, with more serious implications for potential strife.

**The social realm.** The analysis carried out with respect to the political and economic realms must be extended to the social realm as well, but by necessity, the terms of reference vary in this arena. The key issue to be examined with respect to the social realm is the extent of the effective claims made on social esteem and the forms in which these claims translate into differentially distributed but tangible economic and political benefits. Specifically, the examination must focus on whether the given country has status groups based on vocation (i.e., differential standing based on occupation), hereditary charisma (i.e., differential standing based on successful claims of higher-ranking descent), or some version of hierocratic or historical claims (i.e., differential standing based either on religious vocation or specific universally recognized historical actions, or on secular claims associated with national embodiment groups).  In the last category, a common example is the status differential accorded to some groups on the basis of a claim to the national idea, contrasted with groups that may be seen as foreign and having arrived through immigration.  Irrespective of their origins, the size of the status groups needs to be established and the nature and extent of their social claims assessed.  The task provides an understanding of the social profile of the state, including a preliminary judgment about the relative power of various dominant and pariah groups within the country.

Just as in the political and economic realms, a static measure of the size and capabilities of the status group is insufficient to discern the extent and robustness of social closure.  To gauge such dimensions qualitatively, other questions have to be addressed:  How rigid are the extant forms of status stratification?  How controversial are the existing norms of stratification?  What kinds of social mobility are possible, and what is the extent of such mobility?  Are the tangible benefits accruing to members claiming different kinds of social status merely symbolic, or are they substantive and disproportionate relative to the tangible social contributions perceived to be made by the status groups in question?  Focusing on such questions would provide a better sense of how status groups in the social realm intersect with the political and economic structures of power, thereby laying the basis for a clearer idea of the potential for strife.

**Summarizing the notion of closure.**  The overall purpose of the scrutiny of the political, economic, and social system is designed to do three things:  (1) to provide a general profile of which sets of individuals either benefit most or are deprived most by the prevailing social structures when classified by certain substatal categories like race, language, religion, region, or culture; (2) to identify the level of disproportion in the dominance or deprivation of the given group relative to its size or standing in the population as a whole; and (3) to assess the opportunities for peaceful transformation in the system as a whole by means of a careful scrutiny of the existing institutional structures, the ruling norms of behavior, and the general expectations about appropriate political actions.  Taken together, such an examination, conducted skillfully, might suggest whether significant forms of closure exist in any given country.

If significant forms of closure are discovered, it still would be premature to claim that "ethnic" strife is probable.  For such a claim to be sustained, the burdens of closure must be shown to fall disproportionately on individuals with certain potential affinities.  That is, individuals who share possible solidarity deriving from race, language, religion, regional origin, or culture must be either objectively disadvantaged as a result of such closure—whether they know it or not—or they must be capable of being "conscientized" on the basis of some such affinity in the presence of specific perceptions of deprivation.  In such circumstances, it does not matter if the basis for affinity is entirely spurious, for so long as the fact of deprivation (or the appearance thereof) can be shown to fall on individuals potentially linked by "some externally identifiable characteristic,"[36] the potential for usurpation and strife by disadvantaged individuals can be said to exist.  Therefore, the first stage of the model illustrated above is completed by a systematic effort to map the identified patterns of closure onto potential affinities centered either on kinship groups or on other forms of presumed identity based on race, language, religion, geographic origin, or culture.

## STAGE II:  UNDERSTANDING THE TRANSFORMATION FROM POTENTIAL TO LIKELY STRIFE

### Potential Versus Likely Strife

Establishing the potential for strife is not the same as establishing the likelihood of strife.  If objective patterns of closure exist within a given state and if they burden certain sets of individuals with identifiable affinities, the potential for strife certainly exists.  Transforming the potential for strife into likely strife, however, requires certain catalytic elements.  The necessity of a catalyst is particularly salient because large-scale social violence, especially when it involves coordinated actions by numerous individuals, rarely (if ever) occurs as a result of spontaneous social combustion.  The claim that violent

---

[36]Weber, op. cit., p. 342.

action often arises from the perceived "feeling that prevailing conditions limit or hinder"[37] the political prospects of individuals is often rooted in the classical imagery of Jacobin revolutions, which depicted social uprisings as deriving from simultaneous, unconnected, and spontaneous actions by large numbers of discontented individuals. Such a vision received its most sophisticated treatment in the early work of revolution theorists like James Davies and Ted Robert Gurr, who argued that the prospects for large-scale violence are linked primarily to conditions of unbearable immiserization. As Gurr succinctly described the thesis, "the more widespread and intense deprivation is among members of the population, the greater is the magnitude of strife in one or another form."[38] However, neither Gurr nor Davies argues in favor of a simple relationship between deprivation and strife. Both analysts employ intervening variables that center mostly on the capacity of the system to satisfy individual or group expectations. For example, Davies argues that although immiserization sets the stage for revolutionary events, they are most likely to occur "when a prolonged period of objective economic and social development is followed by a short period of sharp reversal."[39] By Davies's logic, such a pattern arises because individual expectations do not decrease in proportion to the state's decrease in institutional capacity; large-scale social violence is a result of rising expectations amidst declining satisfactions. Despite such sophisticated hypothesizing, the core argument still remains fairly simple: relative deprivation, though mediated, constitutes the principal cause of likely strife.

The main problem with such an argument is that it need not be true. To begin with, the deprivation experienced by a given population may not lead to strife or even to opposition if the bonds of solidarity among the affected groups are merely latent or fractured. The latter possibility is particularly relevant because modern societies invariably place individuals in multiple roles. Individuals may acquire multiple identities defined both by the various activities they engage

---

[37]Crane Brinton, *The Anatomy of Revolution*, New York: Harper & Row, 1965, p. 33.

[38]Ted Robert Gurr, "A Causal Model of Civil Strife: A Comparative Analysis Using New Indices," *American Political Science Review*, Vol. 62, 1968, p. 1105.

[39]James Davies, "Toward a Theory of Revolution," *American Sociological Review*, Vol. 6, 1962, p. 5.

in and by the different levels and kinds of meaning bequeathed by their membership in diverse social formations.  There may also be cross-cutting cleavages in society that keep groups from coalescing along certain specific lines despite the existence of otherwise widespread deprivation.  The existence of multiple identities and cross-cutting cleavages can, therefore, conspire to keep relative deprivation from raising the propensity for strife.

Nevertheless, it is important to recognize that deprivational approaches contain an important kernel of truth:  the existence of deprivation is, as Gurr puts it, "the basic condition for civil strife of any kind."[40]  Thus the deprivational approaches are not false, but simply incomplete. They tell only a part of the story.  Their core concerns about inequality, disinheritance, and alienation need to be integrated with other elements that can transform these dispositions into "joint action in pursuit of common ends."[41]  This implies connecting the "diffuse strains and discontents within the population" with a "central political process" so that ethnic-based collective action can be seen for what it actually is:  a species of political action in which "specific claims and counter-claims [are] made on the existing [regime] by various mobilized groups" in order to secure "established places within the [extant] structure of power."[42]  Or, to put it in terms favored by closure theorists, political guidance and organization are needed to transform diffuse dissatisfaction into effective usurpationary action.

The ethnic ingredient by itself never obliterates the basic fact that all such collective actions simply amount to forms of political mobilization designed to capture power or increase existing power.  Consequently, any violence that may accompany mobilization is always political violence, which in turn is "essentially a by-product of [the] omnipresent process of political conflict among mobilized—that is, organized and resource controlling—groups and governments."[43]

---

[40]Gurr, op. cit., p. 1105.

[41]Charles Tilly, *From Mobilization to Revolution*, Reading, PA:  Addison-Wesley, 1978, p. 84.

[42]Charles Tilly, "Does Modernization Breed Revolution?" *Comparative Politics*, Vol. 5, No. 3, 1973, p. 436.

[43]Theda Skocpol, "France, Russia, China:  A Structural Analysis of Social Revolutions," *Comparative Studies in Society and History*, Vol. 18, No. 2, 1976, p. 165.

This argument highlights the complementary dimensions—organized mobilization and reactive state action—necessary to complete the deprivationist narrative and transform it into a coherent theory about ethnic violence because, as Aya remarked, "for all the violent passion and passionate violence they entail, revolutions, rebellion, and lesser forms of coercive civilian conflict are best understood as (to adopt Clausewitz's venerable definition of war) 'a mere continuation of politics by other means.'"[44]

## Mobilization for Political Action

Understanding how individuals who are potentially motivated as a result of deprivation become *mobilized* in the form of "ethnic" groups constitutes the core of Stage II of the model. Five factors are critical to this process.

**Incipient changes in the balance of power.** The first factor consists simply of any significant—incipient—changes in the prevailing patterns of closure or the existing balance of power within a state and sometimes outside it. Since all polities are constantly in flux, thanks to the myriad changes taking place in the political, economic, and social dimensions of public life, social change is clearly as inevitable as it often is mundane. These changes, usually arising out of the normal processes of politics and economics, tend to occur constantly but in incremental form. As such, they rarely garner concentrated attention. Every so often, however, a set of dramatic social changes appear poised to materialize: these changes almost take the form of "step functions" as opposed to the merely accretional modifications constantly witnessed in public life. Changes of this kind can be deliberate, as for example when certain political or constitutional alterations threaten to transform the prevailing patterns of power and privilege; or they can be merely a consequence of an evolutionary process, as for example when certain economic groups finally begin to realize supernormal profits from previously marginally successful ventures; or they can be simply accidental, as for example when the discovery of new resources suddenly increases the wealth of formerly peripheral property owners.

---

[44]Roderick Aya, "Theories of Revolution Reconsidered: Contrasting Models of Collective Violence," *Theory and Society*, Vol. 8, No. 1, 1979, p. 68.

No matter how such changes occur, the critical issue from the point of view of anticipating mobilization is that such changes must be, first, significant in absolute terms and, more important, capable of causing significant alterations in the prevailing internal balances of power.  In other words, the changes worthy of attention are not the mundane changes that constantly occur in social life but the ones that threaten to *immediately* alter the extant patterns of closure in a given society.  The prospect of immediate change is critical from the point of view of mobilization, because social changes that threaten to alter the prevailing patterns of power over the long run—no matter how significant they are and how clearly they are foreseen by the perspicacious—usually do not result in collective action, for two main reasons.  First, individuals tend to discount the future and as such are less motivated to respond to contingencies that do not threaten their immediate interests.  Second, the very length of time required to make long-run changes effective usually implies that many other alternative outcomes could obtain and hence serve to retard pressures for social mobilization and collective action.

Prospects for significant and immediate changes in the prevailing patterns of social closure, however, are perfect precipitants for collective action, and they can work in one of two ways.  To begin with, they can catalyze expectations of quick success and thereby motivate individuals who are dissatisfied with the prevailing balances of power to increase their individual efforts toward group mobilization based on potential solidarities like race, language, regional origin, or culture.  The prospect of incipient change in existing forms of social closure can also work in the opposite direction, in that it can result in collective action by individuals threatened by imminent loss of power or privilege.  While there is no reason—in terms of the principles of standard utility theory—why individuals threatened by loss should behave any differently from individuals in pursuit of gains, there is some experimental evidence to suggest that individuals faced with the prospect of significant, immediate losses tend to be more risk acceptant and, by implication, might be willing to bear greater burdens—if necessary, in the form of enhanced contributions to collective action—in the hope of avoiding such losses.  This evidence, captured in conceptual terms by "prospect theory,"[45] would suggest that fear of immediate loss among the privileged *might* result in more effective collective action—perhaps expressed in ethnic idioms—if such an outcome promises to avert the anticipated losses. The direction in which the pressures toward mobilization run is less important than the factors that cause the pressures. A prospective and consequential change in the prevailing balance of power, even if it comes about purely accidentally, will upset the status quo, creating a disequilibrium that invites collective efforts to either exploit it or ward off its worst consequences.

**Tipping events.** The second factor consists of specific "tipping events" that can galvanize a group into political action. Tipping events are simply any conspicuous public events that arouse group sensibilities, reinforce beliefs in their insular identity, and set off escalating spirals of shared expectations about collective resistance to the established order. By their occurrence, they "confirm or justify the [latent] fears or hatreds in[to] a [more] generalized belief; they may initiate or exaggerate a condition of strain; or they may redefine sharply the conditions of conduciveness [leading to future mobilization and violence]."[46] Tipping events can take various forms. They could include large-scale public violence directed at members of a particular group, the destruction of important properties valued by a community, the forcible relocation, banishment, exile, or execution of important or numerous individuals in a community, or any other such *conspicuous* event that induces or reinforces a differential perception of vulnerability. When viewed in retrospect, tipping events often turn out to be important elements of the historical memory, deepening group identity and accentuating the processes of group formation. They come to serve as the *manifest* symbols of alienation and become "salience indicators"[47] toward which individual fears, resentment, and anger all converge—ready to be molded either by

---

[45]The classic formulation can be found in Daniel Kahneman and Amos Tvesky, "Prospect Theory: An Analysis of Decision Under Risk," *Econometrica*, Vol. 47, No. 2, March 1979, pp. 263–291.

[46]Neil Smelser, *A Theory of Collective Behavior*, New York: Free Press, 1962, p. 17.

[47]The notion of salience indicators was first developed in Thomas Schelling, *The Strategy of Conflict*, New York: Oxford University Press, 1963, p. 98ff.

interested elites or exploited by interested factions both within the state and outside it.

**Leadership.** While "tipping events" can intensify the background conditions for political mobilization and are helpful for that purpose, they are not as essential for ethnic mobilization as the third factor—identity entrepreneurs—usually is. "Identity entrepreneurs"[48] are simply individuals who, for self-interested reasons, find it profitable to contribute to creating group identities and bear the costs of mobilizing such groups for political action. Identity entrepreneurs usually rise from within the ranks of a given substatal affinity, but such attachments are not critical, though probably likely when the mobilization is an ethnic one. These entrepreneurs are generally indispensable for group mobilization because of the problems associated with collective action.[49] Such problems usually derive from the fact that any collective action, whether it be engaging in street protests, mobilizing community action, or orchestrating mass resistance against "outsiders" or the state, is invariably characterized by what has been called "the de facto impossibility of exclusion."[50] This means that it is impossible to prevent the entire set of relevant people from consuming the fruits of a particular collective action, even though only a few may have contributed to their production. For example, a protest action against discrimination mounted by a few individuals may create a political windfall that can be enjoyed by other dominated individuals who did not participate in the original protest. Ethnically driven collective actions are, thus, one more example of such collective goods.

Since the ascriptive tradition of viewing ethnicity regards all ethnic groups as composed ultimately of self-interested, or egoistic, individuals, collective goods—like ethnically based social action—either will not be produced in normal circumstances or at best will be produced only in suboptimal "quantities." The deficit in the production of such goods does not arise because collective ethnic mobilization is

---

[48]The term "identity entrepreneur" has been borrowed from Barbara Ballis Lal, "Identity Entrepreneurs: Do We Want Them/Do We Need Them?" unpublished manuscript.

[49]The *locus classicus* for understanding this issue is Mancur Olson, Jr., *The Logic of Collective Action*, Cambridge: Harvard University Press, 1965.

[50]Russell Hardin, *Collective Action*, Baltimore: Johns Hopkins, 1982, p. 20.

judged to be *unnecessary* in the face of some palpable deprivation, but only because each individual ethnic agent has great incentives to be a "free rider." Knowing that he cannot be excluded from eventually enjoying the fruits of successful ethnic action, he has few incentives to make an individual contribution to its success. This is because he judges his own personal contribution to be either insignificant to the success of the final outcome or not worth the costs he will bear in the process of bringing that outcome about. Consequently, every individual ethnic agent is—in the language of game theory—tempted to "defect" when he has to make good on his individual contribution to collective social action.[51] The effect of every individual agent reasoning this way soon leads to the demise of any possible ethnic action, *even though such action may be both universally recognized as necessary and desired by all individuals sharing certain substatal affinities.*

While it is certainly true that this paradox of "missing" action arises primarily because of a particular view of individual rationality, it is also true that this conception of rationality is sufficiently accurate to describe the motivational structure of most individuals in modern, if not all, societies. Corroborating evidence for this assertion can be seen in the fact that several important collective goods like clean air and public safety cannot be produced except through the intervention of *command* mechanisms like the state. In the case of ethnic mobilization, the state is unavailable as a catalyst and may in fact be an adversary. An alternative mechanism of mobilization is therefore required, and identity entrepreneurs usually fit the bill neatly. Identity entrepreneurs help to resolve the collective-action problems in several ways.[52]

To begin with, they can help to "create" the ethnic group—if the group is merely a latent formation up to that point—by "conscientizing" individuals to support certain collective goals as defined by the entrepreneur. This attempt to define goals for the purpose of mobilizing a given populace is clearly rational and self-

---

[51]Russell Hardin, "Collective Action as an Agreeable *n*-Prisoners' Dilemma," *Behavioral Science,* Vol. 16, 1971, pp. 472–481.

[52]Richard Wagner, "Pressure Groups and Political Entrepreneurs," *Papers in Non-Market Decision-Making,* Vol. 1, 1966, pp. 161–170.

interested behavior by the entrepreneur because its goal is to entice individuals to commit themselves and their resources to political actions that advance the entrepreneur's current and future political interests. However, if individuals are as rational and self-regarding as the ascriptive tradition assumes, they would *not* advance the entrepreneur's interest by participating in such collective action because, other things being equal, they could gamble on being able to enjoy these goods for free after they have been created. Since the costs accruing to each person would certainly be greater than the individual benefits, a person's nonparticipation in the face of exhortations by identity entrepreneurs is just as rational as nonparticipation in the context of other calls for community mobilization—precisely the problem that gave rise to a need for identity entrepreneurs in the first place.

This problem can be solved in several ways. One solution would rely on positing that the identity entrepreneur is a very effective purveyor of "false consciousness." In effect, the political entrepreneur is seen in this instance as exercising great "charismatic leadership" in the Weberian sense. As such, he is credited with the capacity to mobilize sentiments and provide a coherence to the life and actions of individuals in a way that the prevailing social order and the raw fact of latent affinity cannot. The charisma of such leaders serves to suspend the normal calculative rationality of individuals who, putting aside conventional considerations of costs and benefits, tap into their reservoir of extrarational motivations to participate in collective actions despite their potentially high personal cost. While identity entrepreneurs can certainly mobilize collective action in this way by sheer dint of personality, it must be recognized that this process is explicitly extrarational from the ascriptive perspective. It can account for how collective actions originate—activated emotions that almost imply "a fit of absence of mind"—but it is a relatively precarious explanation for why collective actions sustain themselves over time, because charismatic authority is unstable, often transient, and eventually of limited duration simply because of the mortality of individuals.[53]

---

[53]The limitations of charisma as an explanatory concept have been explored in William Spinrod, "Charisma: A Blighted Concept and an Alternative Formula," *Political Science Quarterly*, Vol. 106, No. 2, 1991, pp. 295–311.

An alternative solution to the problem also centers on identity entrepreneurs but runs a more rationalist course. This variant posits that the identity entrepreneur belongs to a "privileged group,"[54] that is, a subgroup with a vested interest in seeing the collective action succeed and willing for purely self-interested reasons to bear the costs of ensuring such success. The self-interested reasons may be political, as is the case when certain individuals perceive that organizing a latent group would advance their own claims to wider political power, or it may be economic, as is the case when certain individuals perceive that empowering a latent group would increase their own markets, wealth, or profits. Whatever the reasons, the identity entrepreneur solves the collective-action problem in this case not by activating the emotions of individuals and gambling on suspended rationality, but by exploiting individual rationality. This is done through the offering of "selective incentives."[55] These may be "positive"—when one or more individuals are compensated (in the form of promises of future political power or simply monetary payments) for their role in orchestrating or participating in collective actions—or they may be "negative"—when one or more individuals are simply coerced into participating in collective actions by threats to their life, liberty, or property. Which sort of selective incentive (or combination thereof) is offered in any given case depends on the character and objectives of the identity entrepreneur in question. What is important, however, is that the mechanism of selective incentives *can* solve the collective-action problem in a direct and rational way by introducing excludability ("those who don't partici-pate, don't get compensated" or "those who don't participate, better watch out") and by equalizing the asymmetry between individual costs and gains ("there are direct benefits for individual participants, not simply diffused gains from future group victories" or "there is special pain for each individual nonparticipant, not simply general discomfort from being quietly worse off").

Because of the solutions they embody, identity entrepreneurs become very important for transforming individuals sharing certain affinities into mobilized political groups pursuing certain power-

---

[54]The phrase "privileged group" comes from Olson, op. cit., pp. 49–52.

[55]Ibid., p. 133.

political goals.[56]  Given the kinds of techniques they can utilize for this purpose, identity entrepreneurs in particular and ethnic political leadership in general can be seen to arise in one or more of three ways.  They may arise from "patrimonial" structures, in which leadership is derived from traditional forms of status and power, both economic and political, as held by certain privileged families; they may arise out of "bureaucratic" institutions, in which leadership is based on special occupational competency and derived from certain training in civil or religious administration or military affairs; or they may arise out of the dynamic personality characteristics and proclivities of an individual.

**Resources and organization.**  Political leadership in the form of identity entrepreneurs is a necessary catalyst but needs to be supplemented by additional components for effective group mobilization.  These additional components are summarized by the fourth factor—resources and organization[57]—in the graphic depicted earlier.  Both elements are critical.  A group's access to resources is in many ways the lifeblood of the mobilization process.  It determines the depth of conscientization that can be undertaken, the level of propaganda and public-relations efforts that must be mounted, the kinds of selective incentives that can be offered to bystanders, and, overall, the kind of strategy that the mobilized group can pursue vis-à-vis other competing social formations and the state.  For all these reasons, identity entrepreneurs arising out of patrimonial structures will invest their own resources in the movement, while those arising out of bureaucratic institutions and personality dispositions will have to rely on the resources of others for sustaining the groups they seek to mobilize.

Access to resources can come in many forms:  besides personal wealth and family connections, which are usually associated with patrimonial structures, identity entrepreneurs can direct their persuasive or coercive efforts at internal community resources, seeking contributions from wealthy ethnic cohorts and appealing to ethnic

---

[56]Douglas A. Van Belle, "Leadership and Collective Action:  The Case of Revolution," *International Studies Quarterly,* Vol. 40, 1996, pp. 107–132.

[57]These factors are most clearly emphasized in the "resource mobilization" perspective in Charles Tilly, *From Mobilization to Revolution,* Reading, PA:  Addison-Wesley, 1978.

émigrés abroad or to foreign states. Success in mobilizing these resources will be determined by the levels of wealth within the group; the confidence individuals have in the existing leadership or the effectiveness of the entrepreneur's coercive threats; the entrepreneur's skill in creating domestic and external coalitions; and the extent and intensity of the deprivation felt by the group at large.

In addition to resources, which provide the lifeblood for effective mobilization, competent organization provides the arteries that channel the lifeblood into effective action. Competent organization is necessary in the presence of every kind of entrepreneurship. Entrepreneurship deriving from patrimonial structures usually brings along some form of nascent organizational structure in the form of patrimonial bureaucracies, but even this organization has to acquire new powers and capabilities if it is to be responsive to the new demands of large-scale political action. Entrepreneurship based on personality disposition or special vocational competency must create organizational structures from scratch or "annex" existing organizational structures which are then oriented to the new goals of ethnic mobilization. These organizational structures are generally needed because communities that possess no hierarchical structure and no functional role differentiation are generally ineffective as agents of *large-scale* social mobilization. For such purposes, a formal organizational structure is required. The structure must be capable of carrying out a variety of actions designed to advance the cause of the mobilized groups and, as such, would have the following three characteristics.

First, it will have a set of closed relationships at the core, meaning that admission to "outsiders" would be proscribed at least at the level of executive authority. The organization may support open structures like a political party, pressure groups, research and educational institutions, and a legal defense arm, but at its core policymaking level, an ethnically mobilized political group is characterized by a closed organization. Second, the organization also will be autonomous and autocephalous, meaning that it frames its own regulations and takes its orders from its own leaders on their personal (or bureaucratically mediated) authority and not from outsiders, no matter how well intentioned; further, it regulates issues of leadership, functional roles, and succession through internal means alone. Third, it will focus special attention on ensuring the security of its leadership and the cadres as well as on preserving the sanctity of the channels of communication linking these entities.

Because political mobilization is oriented to altering the relations of power, the ethnic organization must be attentive to the balance of power among various groups (and the state) as well as to the general perceptions of that balance. Toward that end, it must be able to counter those forces that would oppose continued mobilization and entice those elements that adopt a position of neutrality. Achieving such objectives requires a variety of tactics, the most important of which are successful differential mobilization of resources, intelligence collection on opponent groups, and information warfare. The success of each of these operations is crucially dependent on preserving the security of the leadership-cadre link and the inviolability of the group's means of intelligence gathering, data fusion, and transmission of orders.

**The foreign element.** The final catalyst for effective mobilization of ethnic groups is the possibility of foreign assistance.[58] In a strict sense, foreign aid to a mobilizing ethnic group is simply another facet of the group's effort to muster resources. Because it pushes domestic power struggles into the interstices of interstate competition, however, the role of foreign agencies and states requires separate categorization. The possibility of external assistance to incipiently mobilizing ethnic groups becomes relevant for two reasons. First, state boundaries often do not coincide with the location of ethnic populations. As a result, members of a certain group mobilizing for power within a given country are often tempted to look for assistance from their cohorts located in other states. Such cohorts may occasionally seek to provide tangible support even if it has not been requested by the mobilizing group. Ethnic émigrés abroad are one special case of such possibilities. Second, because interstate competition continues unabated even as domestic struggles for power proceed within states, states may often find it convenient to support domestic challengers abroad simply for purposes of wearing down their competitors. Such support can come in the form of sanc-

---

[58]For a good survey of these issues, see Alexis Heraclides, "Secessionist Minorities and External Involvement," *International Organization*, Vol. 44, 1990, pp. 341–378, and K. M. de Silva and R. J. May (eds.), *Internationalization of Ethnic Conflict*, New York: St. Martin's Press, 1991.

tuary for ethnic elites, financial assistance, diplomatic support, provision of organizational expertise, and even covert arms transfers and military training. Because of the serious and extended nature of such interactions, domestic ethnic mobilization may become swiftly enmeshed in the vicissitudes of international politics. The bottom line, therefore, is that foreign support may be critical for the success of ethnic mobilization, and fears of such support often become critical precipitants of intrastate violence as state authorities are tempted to attempt "preemptive strikes" at an ethnic leadership if consequential foreign support appears imminent or likely. A whole set of issues arises in such situations, as the state authorities then can portray the mobilized ethnic group as a foreign agent and, in turn, claim support from their own allies abroad.

**Summarizing the logic of mobilization.** We examine the factors that can aid or hinder mobilization in order to identify the variables responsible for catalyzing existing latent dissatisfaction into visible political action. When existing differential deprivation is married to the appearance of five factors—the approach of incipient changes, the existence of tipping events, the rise of identity entrepreneurs, the availability of resources and the development of competent organization, and the potential for foreign assistance—the *potential* for strife moves toward *likelihood*. Whether strife actually occurs, however, does not depend just on the capabilities of the mobilized group. The capabilities and actions of the prevailing stakeholders, namely the state, are crucial as well, and it is to these variables that attention must now turn.

## STAGE III:  FROM LIKELY TO ACTUAL STRIFE: UNDERSTANDING STATE CAPABILITIES AND THE PROCESS OF STRATEGIC BARGAINING

### State Capacity

The possibility of actual strife is always determined by the results of the interaction between the preferences, capabilities, and actions of both the mobilized group and the prevailing stakeholders, namely, the state. The state, it is generally assumed, tends to be more powerful (in an absolute sense) than any single challenger. This perception usually derives from the fact that the state is supposed to claim, in popular conceptions, "a monopoly on the means of violence." If this were true in the literal sense of the phrase, the challenge of ethnic strife would be both trivial and uninteresting:  trivial because the state would employ its "monopoly" on force to readily crush any incipient challenge by a substatal group, and uninteresting because all assessments of possible ethnic strife would be little more than a narrative of abortive and failed challenges.

The empirical record, however, suggests that this is not the case. Communal challenges, including ethnic groups, do sometimes succeed, and this is due to the fact that the state—both in theory and in practice—never enjoys a *real* monopoly on the use of force. Weber understood this well enough, and for that reason he described the state as an institution that merely lays *claim* to the monopoly of *legitimate* violence within a territory. Mobilizing groups, therefore, stand a chance when contending with state power because the state, just like other social formations, is constantly struggling to dominate various nodes that generate power within a given society, even as it constantly seeks to increase the levels of coercive power it has available but by no means monopolizes. The mobilized group, then, has certain advantages in that it is never confronted by a "state" that is a single unitary actor with a monopoly on social resources, including the capacity for violence, and therein lie precisely the possibilities for successfully pressing certain political claims.[59]

Understanding how this process can unfold requires an understanding of the nature of the state. The state, simply put, is first and foremost a security-producing *institution*. As such, it involves a territorially defined set of political relationships. These relationships, which take the form of a hierarchy based on coercion and sustained by some concentration of force, are intended primarily for the protection of the dominant elites at the apex but provide, as an unintended consequence, relative safety for the subordinated subjects further down the coercive chain.[60]

---

[59]For an insightful analysis, see Gordon Tullock, *The Social Dilemma: The Economics of War and Revolution,* Blacksburg: University Publications, 1974.

[60]This formulation is amplified and its consequences explored in Ashley J. Tellis, "The Drive to Domination:  Towards a Pure Realist Theory of Politics," unpublished Ph.D. dissertation, The University of Chicago, 1994, pp. 257–285.

This implies that the state, far from being a simple unitary organism, has many "parts." These include the dominant elites at the apex of the institutional structure who "rule" the polity. Below them are several bureaucracies consisting of the army, police, intelligence agencies, tax collectors, and propagandists. The task of such bureaucracies is to maintain security and order with a view to preserving the prevailing structure of power and to effectively collect the revenues required from civil society with a view to both sustaining the structures for power and undertaking redistributional and welfare functions more generally. Finally, there is civil society itself, which consists of ordinary citizens existing atomistically as well as in the form of other large-scale economic organizations like industrial enterprises, agricultural combines, universities, and the like.[61]

Since these formal arrangements mask more invisible patterns of distributed power and wealth, the task of the state essentially boils down to either preserving the existing exclusionary patterns of closure in the polity or preventing the usurpationary attempts at changing existing power relations from manifesting themselves through violence. In effect, therefore, the dominant elites who "rule" the state seek to use the multiple bureaucracies either to preserve their own power or, in more accommodative political systems, to prevent the incipient changes in power relations from taking place through violent means.

Since the ascriptive tradition, however, assumes that all individuals in the system are rational and egoistic, it is evident that the state too faces multiple kinds of collective-action problems as elites attempt to maneuver their underlings to confront the mobilized communal or ethnic groups seeking political change. In the strictest sense, the collective-action problem facing these ruling elites is identical to that facing identity entrepreneurs at the opposite end: to mobilize a group of individuals to undertake a certain task that imposes higher individual costs on each of them in comparison to their individually realized benefits. At the state's end, however, the issue may not be getting individuals to rally behind calls for redistributing certain

---

[61] For a useful survey of the current state of research on the state as a political structure and on the nature of rule, see Theda Skocpol, "Bringing the State Back In: Strategies of Analysis in Current Research," in Peter B. Evans et al. (eds.), *Bringing the State Back In*, Cambridge: Cambridge University Press, 1985, pp. 3–37.

resources—which, for example, is the problem confronting the identity entrepreneur—but rather to find a way to induce numerous citizens to do a variety of individually burdensome tasks, as represented by the challenges facing policemen confronting unruly mobs, armies tasked with hunting down armed resisters, and tax collectors responsible for levying assessments on a recalcitrant populace. In every such instance, the problem remains the same: for the agents undertaking these tasks, the individual burdens are greater than the individual benefits, so each is inclined to avoid contributing in the expectation that he could enjoy all the benefits as long as someone else contributed in his stead.

Precisely to avoid such collective-action problems, state managers do not rely on a collection of atomistic agents but rather structure these agents into a hierarchically ordered organization. Such an organization enables both systematic monitoring of individual action and the efficient coupling of such actions to a structured system of rewards and penalties, thus enabling state managers to extract the desired behavior from what is otherwise merely a large mass of individual actors. The system of rewards and penalties serves to help equalize the disparity between individual costs and benefits facing each agent, though it can never quite equalize these values entirely. For this reason, embedded rules, norms, and expectations of good behavior are developed as part of an organizational "ethos." This organizational ethos serves to provide psychic benefits which, in turn, help to reduce the costs of individual action still further, thus bringing them closer to the level of individually realized benefits.

As an *ultima ratio*, however, action on the part of an individual agent is ensured simply by means of sanctions, which in the extreme may involve the taking of life. This enforcement of sanctions, too, is embodied in a transitive system in which each individual complies with his duties because there is another to punish him, and still another to punish the punisher, and so on. In a simple sense, then, the state is effective (in the sense of being able to compel agents to perform their duties) because the monarch or the ruling elite serve as the "ultimate punishers," as used to be the case in traditional kingdoms ruled by free, physically strong, rulers.[62] As modern political

---

[62]J. R. Lucas, *The Principles of Politics*, Oxford: Clarendon Press, 1985, p. 76.

theory has demonstrated, however, even if such an asymmetrically powerful "omega point" does not exist, agents can still be made to carry out extremely (individually) hazardous orders. Such compliance ensues not because of fear of some omnipotent sovereign (who may not exist), but because each agent, fearing that another (or some others) may obey the existing authority, rushes to obey first. The net result of this logic, in which each agent obeys because of fear that some other(s) might obey, allows state elites to secure the compliance of subordinate agents, despite the fact that these elites may actually be weaker than any given subset of the subordinate population imaginable.[63]

The state, therefore, can solve its collective-action problem, though not assuredly or in every instance. In that sense, it only mirrors the problems faced by an incipiently mobilizing group, which may be successful on some occasions but not others. In any event, however, the capacity of the state will be determined by its ability to successfully undertake not one but two kinds of countermobilization in the face of an emerging ethnic opposition group. First, it has to be able to mobilize its own bureaucracies to undertake a range of direct—immediate—actions against the group, if necessary. Second, and only a trifle less urgently, it must be able to mobilize important sections of civil society, especially critical co-ethnic groups or class formations whose support is essential for the preservation of the existing systems of closure. Here, the state could use its bureaucracies to coerce civil society into providing it with the resources and support necessary for its confrontation with the newly mobilized ethnic group, but such actions do not augur well for the possibility of effectively containing the larger problem.[64]

In any event, a state's capacity to cope with its new "demands groups" is a function of its strength along three dimensions, and each of these must be briefly described as the first step toward understanding how likely strife may be transformed into actual strife. This preliminary step, which consists simply of investigating the

---

[63]For formal proof of this proposition, see Tellis, op. cit., pp. 288–309.

[64]See the discussion in Jeff Goodwin and Theda Skocpol, "Explaining Revolutions in the Contemporary Third World," in Theda Skocpol, *Social Revolutions in the Modern World*, New York: Cambridge University Press, 1994, pp. 28–31.

nature, extent, and depth of state capabilities, is analogous to the previous step of establishing the requirements for successful ethnic mobilization: it completes the story about the required capabilities of the antagonists. Once these state capabilities are explored, it is possible to understand the nature of feasible state responses because this latter set is bounded critically by both the nature of the state's abstract preferences and its existing capabilities.

The first step—understanding state capabilities—therefore requires an examination of the three constitutive facets of state power, namely its accommodative capacity as defined by its political structure, its fiscal capability as measured by the health of its treasury, and its coercive capability, which refers to its ability and willingness to use effective force.

The accommodative capability of the state is probably the most important variable for the purposes of anticipating ethnic violence, since, as argued earlier, ethnic movements are simply special forms of political movements struggling for power. Given this fact, the most important issue for the state is its institutional structure or, in other words, its structured capacity for responsiveness to the demands of its constituents. This variable is, in turn, a function of two specific dimensions: the level of inclusion as expressed by the character of the regime and the nature of organizational design that enables or retards the possibility of peaceful change. The nature of the established political structure, then, combines with the prevailing norms of governance and the cohesion of the ruling elites to determine the accommodative capacity of a given state. This varies as well, and democratic, oligarchic, and authoritarian states will exhibit a variety of different capacities in this regard.

The fiscal capacity of the state is the next important variable for purposes of anticipating ethnic violence because it relates, among other things, to the issue of how a state can ameliorate the demands of mobilized groups short of using force. Clearly, the use of this capacity depends on the character of the political regime to begin with, but it warrants independent analysis because it identifies the margins of maneuver that a state has irrespective of its political structure. Three components are relevant in this regard: (1) the overall condition of the state treasury, meaning the extent of existing surpluses and deficits and the prevailing and prospective composition of budgetary expenditures; (2) the state's potential extractive capacity, meaning its ability to extract additional revenues without fear of accentuating political resistance; and (3) the extent and durability of its social base, meaning the size and wealth of the class base supporting the existing ruling elites.

The coercive capacity of the state is the final variable for the purposes of anticipating ethnic violence because it relates, in the final analysis, to the state's ability to conclusively attempt suppression of political mobilization by force. This variable is ranked third not because of its lesser importance but because coercion is generally understood to be the ultimate arbitrator of rule, though many regimes use coercion quite effectively as the means of first *and* last resort. In any event, the relevant components in this connection are (1) the structural relationship between the ruling elites and the bureaucracies of violence; (2) the social composition of the external and internal security organs; (3) the state's reputation for the use of force, which incorporates historical tradition, attitudes, and experience with respect to the use of force;[65] and (4) the technical capability of the external and internal security forces for purposes of suppressing unrest.

These state capabilities writ large describe in effect its latent capacity to offset popular mobilization by a group challenging the ruling elites currently in power. As such, they represent the state's "assets" (or liabilities, as the case may be) in the interaction with various challenging groups. In contrast to the levels of deprivation and possibilities of mobilization, which represent the "demand" side of the equation relating to probable ethnic violence, the state's capabilities in the accommodative, fiscal, and coercive dimensions then represent the "supply" side.

## Strategic Bargaining

Having completed this examination of state capacity, it is now possible to embark on the second step of Stage III, which consists of

---

[65]Empirical studies have shown a 5–7 year "memory" of democratic and nonrepressive norms that affect negatively the use of force by the state. Christian Davenport, "The Weight of the Past: Exploring Lagged Determinants of Political Repression," *Political Research Quarterly*, Vol. 49, No. 2, June 1996, pp. 377–403.

investigating the nature of the strategic bargaining that ensues between the mobilized group and the state.  This requires systematically matching up the feasible preferences, alternative capabilities, and potential actions of both the mobilized ethnic group and the state to discover the range of social possibilities that could arise as a consequence of such bargaining.  Because this model attempts to assist the analyst tasked with intelligence and warning, and is for that reason fundamentally a conceptual effort at understanding the prospect for ethnic violence long before it materializes, the objective of the following section is not so much to provide accurate "point predictions" of when ethnic violence will occur but rather to structure thought about how alternative combinations of group and state preferences and capabilities interact to produce a variety of political consequences, ranging from political reconciliation to state breakdown.  Because some of these consequences embody strife in varying forms and intensity, they provide a means of "backward deduction," that is, they allow the analyst to discover which combinations of group/state preferences and capabilities are particularly volatile.  If such preferences and capabilities are then seen to materialize empirically in any given case of communal action, the intelligence analysts can "flag" such cases as potentially troublesome.  These cases warrant closer scrutiny by other crisis assessment or "watch" teams, and may even justify preemptive political intervention if these strife-prone areas are of importance to the United States.

Before this process of strategic bargaining is modeled, it is worth mentioning that such a straightforward dyadic interaction between group and state represents only one, and is in fact the simplest, kind of encounter that can be envisaged.  A more complex situation, and perhaps one that has greater empirical fidelity to the real world, would involve not simply one ethnic group facing the state but rather one ethnic group facing perhaps several others in addition to the state.  The origins of such situations are themselves very interesting.  The successful mobilization of one ethnic group may in fact give rise to successful competitive mobilization by other groups, and ethnic competition in such instances may arise not only because the less advantaged group embarks on "usurpationary" closure but also because the more advantaged group feels compelled to preemptively reinforce existing "exclusionary" patterns of closure in order to stave off a challenge to their privileges.  Irrespective of what the immediate causes of ethnic competition may be (or, for that matter, their more remote origins), the fact remains that modeling the processes of strategic bargaining in situations where there are more than two entities remains a very challenging task.

This difficulty is only compounded by the fact that the number of entities in play increases when the state is factored in as a relevant actor.  In fact, in situations where multiple ethnic groups exist in some relationship of super- and subordination among themselves, the relationship of the state to the various mobilized groups becomes a critical factor conditioning the extent of their success.  The degree of autonomy enjoyed by the state from its social bases of power, in conjunction with its own inherent capabilities, will determine whether the state plays the role of an umpire between various contending groups or whether it turns out to be an abettor, if not a direct precipitant, of exclusionary political action.  Modeling the state in each of these roles would be desirable if a complete or "intensive" understanding of the processes of strategic bargaining is required. Such modeling, however, would require formal analysis, including the use of mathematical tools, and lies beyond the scope of this preliminary effort.

Finally, a complete analysis of strategic bargaining would require explicitly incorporating the international system as a causal variable. The international environment not only provides possibilities of external support for mobilized social groups within a state, it also affects the extent and character of *internal* state action through the resources it diverts for external power-political purposes.  In fact, as much modern research has shown, events pertaining to the external environment like victory or defeat in war often act as a precipitant for the mobilization of various social groups within a country, sometimes on ethnic lines.  Consequently, the effect of the external environment—either as efficient or permissive cause—must be incorporated explicitly into an analysis of strategic bargaining for reasons of completeness.

This chapter, however, avoids all the aforementioned complications not because of their lack of importance but because of the need for simplicity.  Since it aims to establish simply "proof of concept" for the intelligence community and other interested observers, it settles for explicating the simplest case:  a dyadic encounter between a single mobilized group and the state.  If the ensuing analysis provides a useful tool for conceptualizing the process of strategic bargaining, it will have served its purpose, while also opening the door to future work (perhaps carried out by others) aimed at "intensively" expanding the model through a formal incorporation of more variables, such as many mobilized groups, the state in multiple roles, and differing opportunities and constraints imposed by the international system.

Given this intention of demarcating the range of outcomes in a simple dyadic encounter between group and state, the process of strategic bargaining is modeled as a three-step process through the means of several tables and matrices.  The steps are linked in such a way that the outcomes of a preceding table or matrix become the dimensions on the axis of the following matrix.

The first step involves "measuring" the capacities of both the mobilized group and the state.  The capacity of the mobilized group is measured with a view to assessing the following:  its ability to be accommodative vis-à-vis other competing social formations, including the state; its ability to sustain the political campaign for redress of its grievances; and its ability to maintain the cohesiveness of its emerging group identity.  Based on the earlier discussion of group formation, the group's capacity to be accommodative, to sustain its political aims, and to maintain its cohesion is measured against binary variations (strong-weak) in (1) leadership, (2) access to resources and organization, and (3) levels of popular support, respectively.

The full extent of the variations in group capacity are depicted in binary terms (high-low) in Table 2.1.

The protocols through which variations in the three categories (leadership, resources, and popular support) lead to specific rankings in the three capacities (accommodative, sustainment, and cohesiveness) are given below.

## Table 2.1

## Capacity of a Mobilized Group

| TYPE OF MOBILIZED GROUP | | CAPACITY | | |
|---|---|---|---|---|
| Code | Descriptors | Accommodative | Sustainment | Cohesiveness |
| A | Strong leadership<br>Good resource support<br>Broad popular support | High | High | High |
| B | Weak leadership<br>Good resource support<br>Broad popular support | Low | High | Low |
| C | Strong leadership<br>Weak resource support<br>Broad popular support | High | Low | High |
| D | Strong leadership<br>Good resource support<br>Weak popular support | Low | High | High |
| E | Weak leadership<br>Weak resource support<br>Broad popular support | Low | Low | Low |
| F | Weak leadership<br>Weak resource support<br>Weak popular support | High | Low | Low |
| G | Strong leadership<br>Weak resource support<br>Weak popular support | Low | Low | Low |
| H | Weak leadership<br>Good resource support<br>Weak popular support | Low | High | Low |

Protocol 1.  Leadership and accommodative capacity linked directly, subject to modification according to popular support values.

1a.  Strong leadership = high accommodative capacity, unless it has weak popular support (in such a case, low accommodative capacity).

1b.  Weak leadership = low accommodative capacity, unless it has weak popular support (in such a case, high accommodative capacity).

Protocol 2.  Resource support and sustainment linked directly, subject to modification according to popular support values.

2a.  Good resource support = high sustainment capacity, unless it has weak popular support (in such a case, low sustainment capacity).

2b.  Weak resource support = low sustainment capacity.

Protocol 3.  Popular support and cohesiveness capacity linked directly, subject to modification according to leadership values.

3a.  Broad popular support = high cohesiveness capacity, unless it has weak leadership (in such a case, low cohesiveness capacity).

3b.  Weak popular support = low cohesiveness capacity.

The capacity of the state is measured in similar (though not identical) areas.  Based on previous discussion, again, the process seeks to assess—specifically—the state's capacity to accommodate the group's demands; its ability to sustain its own political preferences vis-à-vis the mobilized group; and finally, its ability to coerce its opponents in the first or last resort depending on the nature of the state in question.  To assess these capacities, three variables are interrogated in binary terms:  these include the strength of the ruling elites (strong-weak); the vitality of the fiscal base (strong-weak); and the overall character of the political structure (exclusionary-inclusionary).

The full extent of the variations in state capacity are depicted in binary terms (high-low) in Table 2.2.

The protocols through which variations in the three categories (leadership, fiscal position, and type of regime) lead to specific rankings in the three capacities (accommodative, sustainment, and coercive) are given below.

**Table 2.2**

**Capacity of the State**

| TYPE OF STATE | | CAPACITY | | |
|---|---|---|---|---|
| Code | Descriptors | Accommodative | Sustainment | Coercive |
| A | Strong leadership Strong fiscal position Inclusive regime | High | High | Low |
| B | Weak leadership Strong fiscal position Inclusive regime | Low | Low | High |
| C | Strong leadership Weak fiscal position Inclusive regime | High | Low | Low |
| D | Strong leadership Strong fiscal position Exclusive regime | Low | Low | High |
| E | Weak leadership Weak fiscal position Inclusive regime | High | Low | Low |
| F | Weak leadership Weak fiscal position Exclusive regime | Low | Low | Low |
| G | Strong leadership Weak fiscal position Exclusive regime | Low | Low | High |
| H | Weak leadership Strong fiscal position Exclusive regime | Low | High | High |

Protocol 1.  Leadership and accommodative capacity linked directly, subject to modification according to type of regime and fiscal position values.

1a.   Strong leadership = high accommodative capacity, unless it is an exclusive regime (in such a case, low accommodative capacity).

1b.  Weak leadership = low accommodative capacity, unless it is an inclusive regime and a weak fiscal position (in such a case, high accommodative capacity).

Protocol 2.  Fiscal position and sustainment linked directly.

2a.  Strong fiscal position = high sustainment capacity.

2b.  Weak fiscal position = low sustainment capacity.

Protocol 3.  Type of regime and coercive capacity linked directly, subject to modification according to leadership values.

3a.  Inclusive regime = low coercive capacity.

3b.  Exclusive regime = high coercive capacity, unless it has weak leadership (in such a case, low coercive capacity).

The preferences of both group and state are measured, thereafter, on the implicit premise that the nature of an entity's effective capacity would determine its revealed preference.  To be sure, the causal logic may run in the opposite direction in some instances, but such anomalies are not considered because they would be difficult to model (thanks to their inherent indeterminacy) and, more important, because such anomalies would not survive very long if there were a radical disjunction between the preferred objectives/strategies and the underlying capacity to secure/sustain these ends.  On the premise that capacity therefore drives preferences, each entity is allowed a choice of four options:  The mobilized group can choose between negotiation, exploitation, intimidation, or surrender, and the first three choices in every given situation are rank ordered.  The state, in turn, can choose between negotiation, exploitation, repression, or surrender.

Since these strategic choices cannot be identified in the abstract but only in the context of specifically profiled groups facing specifically profiled states, the logical (and probable) choices of both group and state are arranged in matrix form.  That is, the preferences of every conceivable type of group—created as a result of comprehensive variation in the kind of leadership, levels of resources, and extent of popular support (sketched in Table 2.1)—are rank ordered (down to three levels) on the basis of their respective capacities (sketched in

Table 2.1) in the context of hypothetical confrontations with every conceivable type of state.

The protocols governing the preference focus on the leadership capacity as the dynamic driving force behind group actions, though they are subject to modification based on the other two factors. The protocols governing the mobilized group preferences are as follows:

1.  A strong group leadership faced with a strong state leadership will have a preference for negotiation with an inclusive regime.

2.  A strong group leadership faced with a weak state leadership usually will have a preference to exploit or intimidate the latter, depending on other factors (coercive/cohesive values).

3.  A weak group leadership faced with a strong state leadership usually will have a preference for exploitation, but it will opt for intimidation or negotiation, depending on other factors (resources).

4.  A weak group leadership faced with a weak state leadership will almost always opt for exploitation or intimidation, particularly if the former has weak popular support or if the latter is an exclusive regime.

Based on the choices arrived at through such rules, the range of group preferences is demarcated in Matrix 1.

A similar exercise is undertaken for the state. The preferences of every conceivable kind of state—created as a result of comprehensive variation in the kind of leadership, vitality of resource base, and level of inclusiveness in political structure (sketched in Table 2.2)— are rank ordered (down to three levels) on the basis of their respective capacities (sketched in Table 2.2) in the context of hypothetical confrontations with every conceivable type of group.[66]

---

[66]The decision rules for rank ordering derive from an assessment of the relative strength in various dimensions of state capacity—accommodation, sustainment, and coercion—informed by a judgment of state behavior drawn from public choice theory. Since the process of strategic bargaining is part and parcel of the larger processes of political competition, it is logical to assume that securing, holding on to, and eventually augmenting one's political power is the ultimate "prize" of the competition. Consequently, it is reasonable to assume decision rules of the following kind: all state leaders will divest economic power before they divest themselves of political power.

Just as in the case of the group, the protocols governing the preferences focus on leadership capacity as the dynamic driving force behind state actions, though they are subject to modification based on the other two factors.

The protocols governing the state preferences are as follows:

1.  A strong state leadership faced with a strong group leadership usually will have a preference for negotiation unless it is an exclusive regime or the group has weak popular or resource support, in which case it will repress.

2.  A strong state leadership faced with a weak group leadership usually will have a preference to exploit its advantage but may try to repress if the group has weak popular support or to negotiate if it has broad popular support.

3.  A weak state leadership faced with a strong group leadership usually will have a preference for negotiation or exploitation, particularly if the state is inclusive.  If the state is exclusive, it will attempt to exploit or repress before negotiating.

4.  A weak state leadership faced with a weak group leadership will usually try to exploit or repress, unless it is inclusive, in which case it will try to negotiate first.

Based on the choices arrived at through such rules, the range of state preferences is demarcated in Matrix 2.

---

That is, access to wealth is *relatively* less important than determining and enforcing "the rules of the game."  Whether state elites will attempt to coerce first or open the level of access to political rule-making and enforcement will be determined by the nature of the political structures in question.  That is, exclusionary political systems (defined as such both by ethos and institutional design) will attempt to coerce political opponents *before* they give up extant power, whereas more inclusionary political systems will divest control of rule-making and enforcement before they attempt to coerce political opponents.  There is, however, a crucial caveat:  in *all* cases, existing state elites will attempt to coerce political opponents *before* they give up extant power, if the mobilized groups are seen to be beneficiaries of effective foreign assistance, since in all these cases the struggle over internal stratification intersects uncomfortably with the demands of interstate competition.  This latter rule is not reflected in the matrices, but should be kept in mind.

## Matrix 1

## Mobilized Group Preferences

| Mobilized Group Type (Table 2.1) | State Type (Table 2.2) | | | | | | | |
|---|---|---|---|---|---|---|---|---|
| | A | B | C | D | E | F | G | H |
| A | 1. neg<br>2. exp<br>3. int | 1. exp<br>2. int<br>3. neg | 1. neg<br>2. exp<br>3. int | 1. int<br>2. exp<br>3. neg | 1. neg<br>2. exp<br>3. int | 1. exp<br>2. int<br>3. neg | 1. int<br>2. exp<br>3. neg | 1. exp<br>2. int<br>3. neg |
| B | 1. exp<br>2. int<br>3. neg | 1. int<br>2. exp<br>3. neg | 1. neg<br>2. exp<br>3. int | 1. int<br>2. neg<br>3. int | 1. int<br>2. exp<br>3. neg | 1. int<br>2. exp<br>3. neg | 1. exp<br>2. neg<br>3. int | 1. exp<br>2. int<br>3. neg |
| C | 1. neg<br>2. exp<br>3. int | 1. exp<br>2. neg<br>3. int | 1. neg<br>2. exp<br>3. int | 1. int<br>2. neg<br>3. exp | 1. exp<br>2. int<br>3. neg | 1. int<br>2. exp<br>3. neg | 1. int<br>2. exp<br>3. neg | 1. exp<br>2. int<br>3. neg |
| D | 1. neg<br>2. exp<br>3. int | 1. exp<br>2. neg<br>3. int | 1. neg<br>2. exp<br>3. int | 1. int<br>2. exp<br>3. neg | 1. int<br>2. exp<br>3. int | 1. int<br>2. exp<br>3. neg | 1. exp<br>2. neg<br>3. int | 1. exp<br>2. neg<br>3. int |
| E | 1. exp<br>2. neg<br>3. int | 1. exp<br>2. int<br>3. neg | 1. exp<br>2. neg<br>3. int | 1. exp<br>2. neg<br>3. int | 1. exp<br>2. int<br>3. neg | 1. int<br>2. exp<br>3. neg | 1. neg<br>2. exp<br>3. sur | 1. exp<br>2. int<br>3. neg |
| F | 1. neg<br>2. exp<br>3. sur | 1. exp<br>2. neg<br>3. sur | 1. neg<br>2. exp<br>3. int | 1. neg<br>2. exp<br>3. sur | 1. exp<br>2. int<br>3. neg | 1. exp<br>2. int<br>3. neg | 1. exp<br>2. neg<br>3. sur | 1. exp<br>2. int<br>3. neg |
| G | 1. exp<br>2. neg<br>3. int | 1. int<br>2. exp<br>3. neg | 1. neg<br>2. exp<br>3. int | 1. int<br>2. neg<br>3. exp | 1. int<br>2. exp<br>3. neg | 1. int<br>2. exp<br>3. neg | 1. exp<br>2. int<br>3. neg | 1. int<br>2. exp<br>3. neg |
| H | 1. exp<br>2. neg<br>3. sur | 1. exp<br>2. int<br>3. neg | 1. exp<br>2. int<br>3. neg | 1. exp<br>2. int<br>3. neg | 1. int<br>2. exp<br>3. neg | 1. int<br>2. exp<br>3. neg | 1. exp<br>2. int<br>3. neg | 1. exp<br>2. neg<br>3. int |

neg = negotiation; exp = exploitation; int = intimidation; sur = surrender

## Matrix 2

### State Preferences

| Mobilized Group Type (Table 2.1) | State Type (Table 2.2) | | | | | | | |
|---|---|---|---|---|---|---|---|---|
| | A | B | C | D | E | F | G | H |
| A | 1. neg<br>2. exp<br>3. rep | 1. neg<br>2. exp<br>3. rep | 1. neg<br>2. exp<br>3. rep | 1. rep<br>2. neg<br>3. rep | 1. neg<br>2. exp<br>3. sur | 1. rep<br>2. exp<br>3. neg | 1. rep<br>2. exp<br>3. neg | 1. exp<br>2. rep<br>3. neg |
| B | 1. exp<br>2. neg<br>3. rep | 1. neg<br>2. exp<br>3. rep | 1. neg<br>2. exp<br>3. rep | 1. exp<br>2. rep<br>3. neg | 1. neg<br>2. exp<br>3. rep | 1. exp<br>2. rep<br>3. neg | 1. exp<br>2. neg<br>3. rep | 1. exp<br>2. rep<br>3. neg |
| C | 1. neg<br>2. exp<br>3. rep | 1. neg<br>2. exp<br>3. rep | 1. neg<br>2. exp<br>3. rep | 1. rep<br>2. exp<br>3. neg | 1. neg<br>2. exp<br>3. sur | 1. rep<br>2. exp<br>3. neg | 1. rep<br>2. exp<br>3. neg | 1. rep<br>2. exp<br>3. neg |
| D | 1. rep<br>2. neg<br>3. exp | 1. neg<br>2. rep<br>3. exp | 1. rep<br>2. exp<br>3. neg | 1. rep<br>2. exp<br>3. neg | 1. neg<br>2. exp<br>3. rep | 1. rep<br>2. exp<br>3. neg | 1. rep<br>2. exp<br>3. neg | 1. rep<br>2. exp<br>3. neg |
| E | 1. exp<br>2. neg<br>3. rep | 1. neg<br>2. exp<br>3. rep | 1. exp<br>2. neg<br>3. rep | 1. exp<br>2. neg<br>3. rep | 1. neg<br>2. exp<br>3. rep | 1. rep<br>2. exp<br>3. neg | 1. exp<br>2. neg<br>3. rep | 1. rep<br>2. exp<br>3. neg |
| F | 1. rep<br>2. exp<br>3. neg | 1. neg<br>2. exp<br>3. rep | 1. rep<br>2. exp<br>3. neg | 1. rep<br>2. exp<br>3. neg | 1. neg<br>2. rep<br>3. exp | 1. exp<br>2. rep<br>3. neg | 1. rep<br>2. exp<br>3. neg | 1. rep<br>2. exp<br>3. neg |
| G | 1. rep<br>2. neg<br>3. exp | 1. neg<br>2. exp<br>3. rep | 1. rep<br>2. exp<br>3. neg | 1. rep<br>2. exp<br>3. neg | 1. neg<br>2. exp<br>3. rep | 1. rep<br>2. exp<br>3. neg | 1. rep<br>2. exp<br>3. neg | 1. exp<br>2. rep<br>3. neg |
| H | 1. rep<br>2. exp<br>3. neg | 1. neg<br>2. exp<br>3. rep | 1. rep<br>2. exp<br>3. neg | 1. rep<br>2. exp<br>3. neg | 1. neg<br>2. exp<br>3. rep | 1. exp<br>2. rep<br>3. neg | 1. rep<br>2. exp<br>3. neg | 1. exp<br>2. rep<br>3. neg |

neg = negotiation; exp = exploitation; rep = repression; sur = surrender

These two matrices, when viewed synoptically, yield a picture that identifies a wide variety of outcomes, according to a specific group or state type.

The three preferences comprise the likely strategies to be followed by the state and the group against a specific opponent. The choices stem from the specific arrangement of strengths and weaknesses against a specific opponent. The first preference is the primary strategy that either the group or the state is likely to follow. The second preference is the alternative strategy that the group or state may follow. The third choice provides another option that the state or group may consider. The three choices presented here are analytical constructs; in practice the real strategies are likely to be a mixture of the top two or even all three choices, though the top choice generally will be the dominant strategy. For example, if the choices for a state come up as "Negotiate, Repress, Exploit" (in that order), in practice, the strategy of such a state might consist of a sincere negotiating attempt, simultaneously keeping ready a strong capability to crack down on the group (and a readiness to use force, if necessary), and occasional attempts to embarrass or reduce the standing of the group through selective economic or media actions.

Although the preferences do not necessarily imply a temporal dimension, the second strategic preference should be considered by the analyst as having the potential of being followed if the primary strategy is not adopted or is not bearing the desired result. The third preference is included to complete the theoretical picture, but it is likely to be a hedging strategy at best.

Because the final result is always an interaction of the two preferences, a first preference for violence by either the state or the group should be flagged by the analyst as an indication that strife is likely. In cases where both first preferences are for a violent outcome, the potential for violence (perhaps even a very severe type of violence) is high. In cases where neither of the first preferences are for violence, the analyst should look at the second preferences. If one of them is for violence, the case may require monitoring. *A word of caution is in order on using the matrices. The results on violent outcomes that are obtained through the use of the matrices should not be taken as empirical, for they are not. Instead, the results are theoretically justified likely outcomes based on the structure presented above. That is why it is not advisable for the analyst to limit her judgment only to first preferences.*

The suggested format for the analyst is to treat any first preference that includes violence as a "red marker" (violence is likely), any second preference for violence as a "yellow marker" (violence is possible but could be averted), and any third preference for violence as a "green marker" (violence unlikely). The final assessment stems from steps that link evidence to specific group and state types, and which in turn lead to a variety of outcomes.

Since the propensity for violence, however, is the specific issue of interest here, the state and group types that yield violent preferences most often should be a special concern for the intelligence analyst. An examination of the results (first and second preferences matched together) point to some insights (see Table 2.3).

Other than the linkage to preference for violence among states that have exclusionary political structures,[67] there is no clear pattern in the rankings.

## Table 2.3

### Characteristics of States and Groups, Ranked According to Their Propensity Toward Violence
(First + second preference; total set $n + n$ is 8 + 8)

State types
| | |
|---|---|
| D = 6+1 | Strong leadership, strong fiscal position, exclusive regime |
| G = 6+0 | Strong leadership, weak fiscal position, exclusive regime |
| F = 5+3 | Weak leadership, weak fiscal position, exclusive regime |
| H = 4+4 | Weak leadership, strong fiscal position, exclusive regime |
| A = 4+0 | Strong leadership, strong fiscal position, inclusive regime |
| C = 4+0 | Strong leadership, weak fiscal position, inclusive regime |
| E = 0+1 | Weak leadership, weak fiscal position, inclusive regime |
| B = 0+1 | Weak leadership, strong fiscal position, inclusive regime |

Group types
| | |
|---|---|
| G = 5+1 | Strong leadership, weak resources, weak popular support |
| B = 4+2 | Weak leadership, good resources, broad popular support |
| F = 3+3 | Weak leadership, weak resources, weak popular support |
| C = 3+2 | Strong leadership, weak resources, broad popular support |
| D = 3+0 | Strong leadership, good resources, weak popular support |
| H = 2+4 | Weak leadership, good resources, weak popular support |
| A = 2+3 | Strong leadership, good resources, broad popular support |
| E = 1+3 | Weak leadership, weak resources, broad popular support |

---

[67]This finding has been borne out in empirical studies; see Rudolph J. Rummel, "Is Collective Violence Correlated with Social Pluralism?" *Journal of Peace Research,* Vol. 34, No. 2, 1997, pp. 163–175.

Another way of portraying the propensity of group and state types for violent outcomes is to show the correlation of the first preference for a violent outcome that each group and state type has in a matchup with all other state and group types, respectively, and regardless of whether it is a preference of the state or the group. In other words, the calculation includes (1) taking, for example, state type A and checking its first preferences against all group types; (2) taking all group types and checking their preferences versus state type A. Table 2.4 shows how the various states and groups come out on such a scale.

Comparison of the two tables shows some shifts in position but only one major shift, as group D moves to near the top of propensity for violence among all groups. The different position of group D on the two tables indicates that even though the group-D type itself is not prone toward violence, states tend to prefer violence in dealing with it.

Table 2.4

**Propensity of States and Groups Toward Violence**
**(First preference, either by group or by state; shown as percentage,**
**total set ($n$) for each is 16)**

State types
| | |
|---|---|
| F = 69 | Weak leadership, weak fiscal position, exclusive regime |
| D = 69 | Strong leadership, strong fiscal position, exclusive regime |
| G = 50 | Strong leadership, weak fiscal position, exclusive regime |
| H = 31 | Weak leadership, strong fiscal position, exclusive regime |
| E = 25 | Weak leadership, weak fiscal position, inclusive regime |
| A = 25 | Strong leadership, strong fiscal position, inclusive regime |
| C = 25 | Strong leadership, weak fiscal position, inclusive regime |
| B = 13 | Weak leadership, strong fiscal position, inclusive regime |

Group types
| | |
|---|---|
| G = 63 | Strong leadership, weak resources, weak popular support |
| D = 56 | Strong leadership, good resources, weak popular support |
| F = 50 | Weak leadership, weak resources, weak popular support |
| C = 44 | Strong leadership, weak resources, broad popular support |
| H = 38 | Weak leadership, good resources, weak popular support |
| A = 31 | Strong leadership, good resources, broad popular support |
| B = 25 | Weak leadership, good resources, broad popular support |
| E = 19 | Weak leadership, weak resources, broad popular support |

This research effort has made no attempt to identify all possible outcomes that can be derived by varying the decision rules or integrating ambiguity in the form of nonbinary choices like "neither strong nor weak" or "medium." Incorporating such choices makes the outcomes themselves more ambiguous. This may in fact be a more accurate reflection of what social events look like in practice, but such an effort is not justified by the intentions underlying this theoretical effort. After all, the objective here is not to provide a "ready reckoner" that can render irrelevant the knowledge or judgment of regional analysts, but rather to provide a heuristic device that frames both the understanding of communal mobilization and identifies the kinds of variables that may have a bearing on the choice for violence by the mobilized group or the state. To that degree, this theoretical model must be viewed as a "first cut," an invitation to further research both conceptual and empirical as well as a template that can be further elaborated and embellished.

# Chapter Three

# THE YUGOSLAV RETROSPECTIVE CASE

## Thomas S. Szayna and Michele Zanini

## INTRODUCTION

This chapter applies the "process" model for anticipating the incidence of ethnic conflict to the case of the collapse of the federal Yugoslav state in 1991. It examines the case of Yugoslavia from the perspective of what an analyst might have concluded about Yugoslavia's propensity toward ethnic violence had she used the "process" model to examine the situation in Yugoslavia in late 1989 or early 1990.

In essence, this chapter examines the ethnic mobilization of Serbs in Yugoslavia under the leadership of Slobodan Milosevic and this group's attempt to alter the federal setup of Yugoslavia in favor of the ethnic Serbs. As such, the conflict in question is between the mobilized ethnic Serbs and the federal Yugoslav state. The primary time frame of interest is the period 1986–1989. The orientation is in line with the focus of the "process" model on one particular kind of ethnic conflict, namely, the rise of an ethnic group challenging the state.

The choice of the date stems from two reasons. One, for purposes of retroactive validation of the model, the cutoff date offers enough data on the strength of ethnic Serbian mobilization. An earlier cutoff date, for example mid-1987, would not have led to different results but would have prevented the inclusion of several factors that aided the mobilization process. Two, the time frame is realistic in that the fate of Yugoslavia (collapse into violent civil war) still was not preordained in late 1989. Many factors, internal and external, still might have headed off the conflict. By early 1991, the writing was on the wall, but even by January 1990, there was nothing inevitable about the breakup.

The retrospective application of the model that is presented in this chapter is far from a trivial exercise that goes over ground already well covered in the numerous analyses of the Yugoslav breakup. We attempt here to walk the reader back to the late 1980s and the euphoria over the end of the Cold War and the wave of democratization sweeping central and southeastern Europe. At the time, U.S. intelligence estimates warned that a breakup of Yugoslavia was likely to be violent, but the likelihood of lengthy, widespread, and brutal internecine armed strife within Yugoslavia still seemed far-fetched to most observers and analysts. Yugoslavia had survived for decades, and it made little sense that one of the most prosperous and free of the European communist states would go down the path of spiraling violence and self-destruction. Fitting in with the general perception of post-Tito Yugoslavia as a loose and unwieldy entity, most observers expected even a greater loosening of the federal setup, rancorous negotiations between the various entities of the federal state, and a "muddling through" the economic and political problems. Could we have done better in forecasting the collapse of Yugoslavia? This chapter takes on that question.

As presented in the model, the group-versus-state conflict is the simplest form of ethnic competition and, some may argue, not realistic when applied to the multiethnic conditions of 1980s Yugoslavia. A model focusing on the intergroup ethnic competition in Yugoslavia—for example, between Serbs on the one hand and Slovenes and Croats (in slightly different roles) on the other, with the federal Yugoslav state playing the role of umpire—might provide an alternative tool to examine the situation. But such an analysis seems overly tainted by the benefit of hindsight and focuses more on the situation in 1990–1991 just prior to the outbreak of violence, rather than on the initial challenge to the constitutional setup. The Serb mobilization led to countermobilizations along ethnic lines in some of the other Yugoslav republics and certainly cemented the demise of federal Yugoslavia, but the Serb mobilization was the first and most important cause of the destruction. The various ethnic groups inhabiting Yugoslavia had numerous grievances against the federal government, but none of the major ethnic groups or republics had mobilized on a secessionist platform until the rise of militant Serbian nationalism destroyed the federation and caused them to opt out of the federal setup. Even in early 1991, the Croat leadership still thought in terms of an accommodation within the bounds of a looser—but still united—Yugoslavia. In that sense, ethnic mobilization of the Serbs against the existing federal structure was the prime cause for the breakup of the country.

Following this introduction, the chapter has four sections. First, it examines the structure of closure according to the questions outlined in the model. That section provides an analysis of which ethnic groups were privileged and which were dominated (the demographic characteristics of Yugoslavia in the 1980s, on which the analysis is based, is appended at the end of the chapter). Then it examines the strength of the challenging ethnic group—the ethnic Serbs of Yugoslavia—by looking at its mobilization process within the categories outlined by the model. An analyst looking at Yugoslavia in the late 1980s would have been wise to examine all of the major ethnic groups in Yugoslavia from such a perspective, but for reasons of space, the desire to avoid duplication, as well as *ex post facto* knowledge, this examination is limited to the Serbs. The second section looks at the capabilities the state—federal Yugoslavia—could have brought to bear in dealing with the challenging group. The third section examines the strategic choices, arrived at on the basis of the assessments in earlier sections, that the state and the group were likely to pursue vis-à-vis each other, given their resource base group. The fourth—and final—section contains some observations on the applicability of the model. All the data used here come from publicly available sources and would have been accessible to analysts in the late 1980s. Wherever data are unavailable, an educated guess is postulated and the reasoning behind it explained.

## ASSESSING THE POTENTIAL FOR STRIFE

### Closure in the Political and Security Realms

In terms of closure in the political realm, centers of political power in Yugoslavia were located at both the federal and republican levels in an elaborate system of shared governance and checks and balances, primarily by the republics upon the federal government. The principle of strict balancing of top political authorities (akin to quotas) by republic was present at all levels of Yugoslav institutions, most visible at the highest levels of political power. Rotating or short-term chairmanships (often apportioned on the basis of observing ethnic and republican balance) of all top bodies was the norm.

The ethnic and republican criteria for representation on all governmental bodies prevented any group from dominating the Yugoslav political realm. However, an ethnic Serb usually held one of the top posts in Yugoslav federal political institutions. The phenomenon stemmed more from the fact that ethnic Serbs from a variety of republics (Serbia proper, Croatia, Bosnia-Herzegovina) or provinces (Kosovo, Vojvodina) could be representatives of that region to the specific federal body. Unlike the Serbs, few members of other ethnic groups lived outside their eponymous republics, causing their ethnic representation at the Yugoslav level to occur almost exclusively within their republican quota. The quota system was based on Article 242 of the 1974 constitution, which endorsed this "nationality key" policy but did so on the basis of "republican" rather than "national" proportional representation. However, with the exception of the Serbs, republican and national representation were functional equivalents.

These patterns are evident from a more detailed look at the composition of personnel at the upper levels of the political apparatus in Yugoslavia. In terms of the highest-ranking individuals, the Yugoslav Presidency, the Federal Executive Council, the Federal Assembly, and the League of Communists of Yugoslavia (or LCY, the name of the communist party of Yugoslavia) were the most influential institutions at the federal level. Due to rotating short-term chairmanships, the larger membership of the bodies (as opposed to the official serving as leader at any particular time) is shown in summary tables below, along with each group's ethnic composition.

The Presidency was the collective head of state. The 1974 constitution provided for the Presidency having nine members, consisting of one representative from each of the republics and autonomous provinces and the chairman of the Presidium of the LCY. The composition of the Presidency changed in 1989, with the number lowered to eight through the removal of LCY representation. The president and a vice president were appointed from the presidency group for a term of one year. The president of the Presidency rotated yearly to

provide even distribution among the republican and provincial representatives.[1]  Table 3.1 summarizes the specific division of posts
within the presidency at the beginning of 1990 by ethnicity.

The Federal Executive Council (FEC) was the executive body of the
Federal Assembly (parliament).  It included a prime minister (FEC
president), two deputy prime ministers and twelve secretaries in
charge of an equal number of secretariats (equivalent to ministries).
In the 1980s, the single most important political post in the federal
Yugoslav government was probably that of prime minister.  Although
the practice was not always followed to the letter, the FEC maintained an affirmative action scheme based on "nationality" as a way
to allocate its senior posts.  It also added ministers without portfolio
from those republics that were underrepresented in ministerial
posts.  Table 3.2 shows the breakdown of important postholders by
ethnicity in early 1990.  While specific snapshots of this type may be
deceiving, the political deals that underpinned the distribution of
ministerial posts remained relatively constant.  Thus, throughout the

### Table 3.1

### Yugoslav State Presidency, January 1990

| Position | Name | Ethnicity |
| --- | --- | --- |
| President | Janez Drnovsek | Slovene |
| Vice President | Borisav Jovic | Serb |
| Members | | |
| Bosnia-Herzegovina | Bogic Bogicevic | Serb |
| Croatia | Stipe Suvar | Croat |
| Kosovo | Riza Sapundziju | Albanian |
| Macedonia | Milan Pancevski | Macedonian |
| Montenegro | Nenad Bucin | Montenegrin |
| Serbia | Borisav Jovic | Serb |
| Slovenia | Janez Drnovsek | Slovene |
| Vojvodina | Dragutin Zelenovic | Serb |

SOURCE:  CIA, *Directory of Yugoslav Officials,* March 1990.

---

[1]Beginning with the 1989 president, Janez Drnovsek of Slovenia, the presidency of the
Presidency was to rotate among the republics and provinces in the following order:
Serbia, Croatia, Montenegro, Vojvodina, Kosovo, Macedonia, and Bosnia-Herzegov-
ina, through 1997.

Table 3.2

Federal Executive Council, January 1990

| Position | Name | Ethnicity |
|---|---|---|
| President | Ante Markovic | Croat |
| Vice Presidents | Aleksandar Mitrovic | Serb |
| | Zivko Pregl | Slovene |
| Ministry | | |
| Internal Affairs | Petar Gracanin | Serb |
| National Defense | Veljko Kadijevic | Serb |
| Justice | Vlado Kambovski | Macedonian |
| Foreign Affairs | Budimir Loncar | Croat |
| Finance | Branko Zekan | Croat |

SOURCE: CIA, *Directory of Yugoslav Officials,* March 1990.

1980s, at any one time, several Serbs were on the FEC and a Serb was one of the vice presidents (Croats occupied the post of president of the FEC in the period leading up to the breakup).

The 1974 constitution divided the Federal Assembly into two chambers, the Federal Chamber (220 delegates, with each republic and province having 30 and 20 delegates, respectively) and the Chamber of Republics and Provinces (88 delegates, with each republic and province having 12 and 8 delegates, respectively). Table 3.3 displays the ethnic composition of the Federal Assembly presidency and commissions at the beginning of 1990.

Table 3.3

Yugoslav Federal Assembly, January 1990

| Position | Name | Ethnicity |
|---|---|---|
| President | Slobodan Gligorijevic | Serb |
| Vice President | Suada Muminagic | Muslim* |
| Secretary General | Ljubomir Bulatovic | Bosnia (Serb)* |

*Probable ethnicity, though reliable data are unavailable.
SOURCE: CIA, *Directory of Yugoslav Officials,* March 1990.

While the Federal Chamber's function was to ensure representation of grassroots organizations at the federal level, its delegates would typically follow instructions from their republican governments Table 3.4 shows the distribution of top posts within the body at the beginning of 1990.

The LCY Presidium (equivalent to the Politburo in other communist states) had the function of a steering body for the Central Committee of LCY and provided leadership for the party between its congresses. The Presidium included a host of commissions to monitor and implement party policy. Like other Yugoslav institutions, the LCY adopted an elaborate quota system: for instance, the chairmanship of the Presidium followed a "nationality key" that ensured periodic rotation of the post among the six republics and the two autonomous provinces. Table 3.5 presents the top posts within the party by ethnicity in January 1990 (the Slovene representative is not listed in the table, since the Slovene party delegation withdrew from the LCY during the January 1990 party congress).

The institutional setup at the level of the republics and the autonomous provinces mirrored the federal one. In multiethnic republics such as Bosnia-Herzegovina, affirmative action schemes (bordering on quotas) were in place to ensure a balanced representation of different nationalities

The information presented above focuses on the highest-level individuals. Less information is available on the ethnic breakdown of the

### Table 3.4

### Yugoslav Federal Chamber, January 1990

| Position | Name | Ethnicity |
| --- | --- | --- |
| President | Bogdana Glumac-Levakov | Serb* (Vojvodina) |
| Vice President | Lazo Tesla | Croat* |
| Secretary | Aleksandar Vujn | Serb* |

*Probable ethnicity, though reliable data are unavailable.
SOURCE: CIA, *Directory of Yugoslav Officials,* March 1990.

Table 3.5

LCY Presidium, January 1990

| Position | Name | Ethnicity |
|---|---|---|
| President | Milan Pancevski | Macedonian |
| Secretary | Petar Skundric | Serb |
| Members | | |
| Bosnia-Herzegovina | Ivan Brigic | Croat |
| Croatia | Marko Lolic | Serb |
| Montenegro | Miomir Grbovic | Montenegrin* |
| Macedonia | Milan Pancevski | Macedonian |
| Serbia | Petar Skundric | Serb |
| Bosnia-Herzegovina | Ugljesa Uzelac | Muslim* |
| Macedonia | Ljubomir Varoslija | Macedonian* |

*Probable ethnicity, though reliable data are unavailable.
SOURCE:  CIA, *Directory of Yugoslav Officials*, March 1990.

upper staff of the bureaucracies led by these individuals. Neverthe-less, by law and custom, the patterns of republic and ethnic balanc-ing in all federal institutions and bodies were present in the appor-tionment of managerial spots. Available information points to enforced ethnic and republican "affirmative action" in selection of personnel. Thus, it is unlikely that any major deviations from the rule of strict ethnic balancing were present through the late 1980s.

The replication of the ethnic balancing took place also in institutions at the republican level, leading to patterns of representation roughly similar to the ethnic balance in the republic. However, Serbs were often overrepresented in both republican and federal institutions, especially in the LCY. By the early 1980s, 47 percent of all LCY mem-bers were Serbs; Serbs were also overrepresented in the ranks of the communist party in Croatia (around 35 percent of total membership) and Bosnia-Herzegovina (47 percent).[2]

In terms of closure in the security realm, federal laws called for a system of representation within the armed forces similar to that in

---

[2]V. P. Gagnon, "Ethnic Nationalism and International Conflict: The Case of Serbia," *International Security*, Vol. 19, No. 3, Winter 1994–95, p. 149.

the country as a whole.  In practice, however, these guidelines were not observed and Serbs served in a dominant role in the military and police apparati.  The phenomenon may have stemmed as much from self-selection as from ingrained institutional favoritism toward one ethnic group.  But members of other ethnic groups aspiring for careers in the military or security bureaucracies seem to have had "reserved" slots in secondary career paths within these bureaucracies.  A more detailed look at the ethnic composition of security apparatus personnel follows.

At the federal level, the Secretariat for National Defense and the Secretariat for Internal Affairs were the prime centers of influence in security matters.  The armed forces consisted of the Yugoslav People's Army (JNA) and the Territorial Defense Forces (or TDF, a large militia force with territorially organized units throughout the country).  While the Presidency was entrusted with command of the armed forces, the secretary for national defense held operational control over the JNA.  The chief of the JNA General Staff served as a deputy to the secretary for national defense; often, chiefs of the General Staff would be promoted to secretaries for national defense Tables 3.6 and 3.7 provide the ethnic background of the top defense and military leaders in early 1990.  The minister of defense, General Kadijevic, had a Croat-Serb background but self-identified as a pro-Yugoslav Serb.[3]

### Table 3.6

### The Federal Secretariat for National Defense, January 1990

| Position | Name | Ethnicity |
| --- | --- | --- |
| Federal Secretary | Veljko Dusan Kadijevic | Serb |
| Deputy Secretary | Stane Brovet | Slovene |

SOURCE: CIA, *Directory of Yugoslav Officials,* March 1990.

---

[3]General Veljko Kadijevic, *Moje Vidanje Raspada: Vojska bez Drsave* (My View of the Breakup: Army Without a State), Belgrade:  Politika, 1993.

Table 3.7

Yugoslav People's Army, January 1990

| Position | Name | Ethnicity |
|---|---|---|
| Chief of Staff | Blagoje Adzic | Serb |
| Deputy Chiefs | | |
|   Air Force | Nikola Maravic | Croat* |
|   Ground Forces | Dragisa Drljevic | Montenegrin* |
|   Intelligence | Djordje Mirazic | Slovene |
|   Navy | Vjekoslav Culci | Serb* |
|   TDF | Ilija Boric | Croat* |

*Probable ethnicity, though reliable data are unavailable.
SOURCE:  CIA, *Directory of Yugoslav Officials,* March 1990.

The Federal Secretariat for Internal Affairs regulated the work of republican and provincial secretariats for internal affairs, including certain aspects of the judicial system (such as the work of public prosecutors).  The secretariat often played a key role in ensuring public order, and it availed itself of the State Security Service (SDB) and the People's Militia.  The SDB was a secret police network tasked with the neutralization of "enemies of the constitutional order"; the People's Militia was a well-trained and equipped paramilitary force. Secretaries for internal affairs at the republican and provincial level controlled their own police forces.  Table 3.8 provides the ethnic background of the top internal affairs officials in early 1990.

The ethnic makeup among the individuals who formed the upper levels of the security and armed forces apparati appears to have been skewed in favor of Serbs and Montenegrins.  The pattern was evident in the armed forces, and we assume it was replicated in the federal police apparatus.  The predominant Serb presence in the armed forces happened despite the constitutionally mandated proportional representation of JNA enlisted and officer ranks according to nationality.  The "nationality key" affirmative action system was supposed to be binding in the armed forces, from the rank of colonel and above, just as in the case of other federal institutions.

By 1983, Serbs constituted almost 60 percent of the officer corps, with an even greater presence in the high command positions.

**Table 3.8**

**Secretariat for Internal Affairs, January 1990**

| Position | Name | Ethnicity |
|----------|------|-----------|
| Federal Secretary | Petar Gracanin | Serb |
| Deputy Secretary | Slobodan Tradijan | Serb* |
| Under Secretary | Zdravko Mustac | Croat |
| SDB Chief | Zdravko Mustac | Croat |

*Probable ethnicity, though reliable data are unavailable.
SOURCE:  CIA, *Directory of Yugoslav Officials*, March 1990.

Nearly every national defense secretary was a Serb (or Croats from a Partisan background). Montenegrins were also overrepresented at the officer corps level, making up 10 percent of its ranks (Montenegrins constituted 3 percent of the Yugoslav population). Croats and Slovenes were the most underrepresented in the JNA officer corps, making up 15 percent and 5 percent, respectively, of its ranks. Muslims, Albanians, Macedonians, and Hungarians made up only a small percentage of the officer corps. The domination of Serbs in the JNA was not uniform across services. While Serbs were disproportionately represented in the ground forces (the dominant branch of the JNA), the air force and navy had substantially more officers from other ethnic groups in their leadership.[4]

The underrepresentation of Croats and Albanians in the JNA may have stemmed from informal restrictions. For example, the reduced Croat presence was in part due to the purge of alleged Croat separatist officers from the JNA in the early 1970s. Distrust of ethnic Croats seems to have remained among Serb officers, and the representation of Croats in the JNA never recovered to their previous levels.

**Assessment of closure in the political and security realms.** The static "snapshot" presented above portrays vividly the effect of the whole system of rules and customs set up to ensure that no single ethnic group would be able to "capture" all (or even the majority) of

---

[4]David Isby, "Yugoslavia 1991—Armed Forces in Conflict," *Jane's Intelligence Review*, September 1991, p. 397.

the top political posts at the federal level. Similar systems of rules and customs operated at the republican level in the multiethnic republics, such as Bosnia-Herzegovina. Despite the presence of a balancing system, Serbs appear to have been overrepresented in the lower levels of the federal bureaucracies and especially in the LCY— the gateway to all positions of political power.

The overrepresentation of Serbs in Yugoslav political structures was not necessarily caused by a deliberate attempt to exclude members of other ethnic groups. Other significant factors may have played a role, including the fact that federal jobs were not considered as prestigious and were not sought after by residents outside Serbia as within Serbia or among Serbs in general.[5] Moreover, constitutional provisions mandated that quotas be determined not by ethnicity but by republic. Thus, Serbs were overrepresented at the federal level in part because they were overrepresented in republican institutions in Croatia and Bosnia-Herzegovina. The overrepresentation at the republican level in turn was partly caused by the fact that Serbs made up a disproportionate amount of the LCY membership.

There were some differences between the political and security realms. Most of the federal political institutions maintained a rough ethnic balance at the elite and director level at the beginning of 1990. Other than the oligarchic function of the LCY in limiting access to top political posts, there appears not to have been any major noticeable patterns that would point to closure along ethnic lines in the political realm. The extensive rules on republican and ethnic representation in all federal bodies were largely observed.

The rank-and-file membership of the military and security apparati differed little from the general patterns within Yugoslavia. However, discrepancies appeared at the mid-to-high levels. Within the confines of this pattern, Serbs and Montenegrins were overrepresented, and most of the other ethnic groups were underrepresented (especially Croats, Slovenes, and Albanians). At the highest levels, it was politically important to fill the most prestigious seats of power in a representative fashion along ethnic lines.

---

[5]Susan L. Woodward, *Balkan Tragedy: Chaos and Dissolution After the Cold War*, Washington, D.C.: The Brookings Institution, 1995, p. 109.

Was there a mechanism that allowed for change of the "snapshot" presented above?  Theoretically, no formal restrictions on access to political power existed, and the government built an elaborate system of rules to prevent such restrictions.  However, membership in the LCY was formally necessary for promotion beyond a certain level of responsibility.  The LCY jealously guarded its special role as the only channel to political power, and it never hesitated to crack down on any signs of a challenge or dissent.  The attractiveness of the LCY primarily to Serbs (and Montenegrins) on the one hand, and its relative lack of attractiveness to Croats and Slovenes on the other hand, acted as an indirect filter of personnel.

The flip side of the emphasis on ethnic representation and balancing was that, due to imbalances in candidate availability and quality by ethnic group, merit sometimes became secondary to ethnic criteria for advancement.  The result was that both minority groups and any major group that was already overrepresented in the federal bureaucracy (especially the Serbs) sometimes were disadvantaged.  To illustrate this point and to put it in terms of a simple matrix, each candidate for a promotion could be defined as either majority/minority and high/low quality.  Sometimes a majority candidate of high quality was passed over for promotion in favor of a minority candidate of high quality.  But in other cases a majority candidate of high quality might have been passed over in favor of a minority candidate of low quality (because of the ethnic balancing principle).  In both cases, the majority candidate who did not receive the promotion might have become resentful, blaming only her majority status for losing out on advancement (though she would have been fully correct only in the second instance).  The inverse of the above was also problematic.  For example, a minority candidate of low quality might have been passed over in the promotion process in favor of a majority candidate of high quality.  In other cases, a minority candidate of low quality might have been passed over in favor of a majority candidate of low quality (for example, due to a specific political deal or because of the winner's personal connections and influence with the decisionmakers).  In both cases, the minority candidate who lost out might have become resentful, blaming only her minority status for not advancing (though she may have been correct only in the second instance).  Such dynamics can be worrisome in conditions of only two ethnic groups.  But in Yugoslavia, with its eight major ethnic groups, a dozen minor groups, and different interrepublican and interprovince population characteristics (and with the ethnic balancing schemes applied both at the federal level and the republic and province levels), using ethnic criteria as a major consideration for promotion and advancement was potentially highly divisive.

Those overrepresented in the pool of candidates (LCY), most of all the Serbs, faced relatively greater competition (among themselves) for fewer seats than, for example, Croats, whose pool of qualified candidates was smaller (because of lower LCY membership). Whereas ten Serbs may have been eligible for one high-level security position, only three Croats may have been eligible for a similar position. But if one Croat and one Serb were chosen from the pools of different size, the Serbs not chosen might have been resentful of the "lower-qualified" Croat being promoted over them.

However, the true losers of the ethnic balancing process were the small ethnic groups (smaller than the eight main groups) who did not "matter enough" in the larger political deals. With top spots virtually reserved for members of the major ethnic groups (in order to adhere to the principles of ethnic proportionality), members of the smaller groups faced an informal ceiling on how far they could advance. For example, with the top positions apportioned on the basis of ethnicity and candidates for these positions judged more on ethnicity than on the basis of merit, it was difficult for, say, a Bulgarian from southeastern Serbia to be appointed to the spot reserved for Serbs, for which Serbs believed they already had a pool of qualified Serb candidates. Perhaps most paradoxically, those who self-identified as "Yugoslavs" (usually the offspring of ethnically mixed marriages) also faced difficulties in advancement, since the categorization scheme in place in Yugoslavia did not envision any quotas for those who considered themselves "Yugoslavs." In any event, in the political realm an informal hierarchy developed, with Serbs at the top of the hierarchy (most numerous and eligible to represent several republics or provinces), then the other groups having a republic status, then the groups having a province status, and then the others.

There was little potential for peaceful change of the informal ethnic constraints on access to political power. The collectivist principle of equal rights for all ethnic groups was subordinated to the general communist political system of rule in the country. Throughout the post–World War II period, elite positions in government were reserved for LCY members. Socialization within the LCY was meant to ensure acceptance of the overall political setup. Suppression of dissent to the principle of LCY leadership was a consistent feature of the Yugoslav regime. Under the authoritarian aspects of the political system as a whole, ethnicity per se was not a bar to advancement, though the way that ethnicity crept in as a criterion for promotion affected different groups in different ways.

Within the constraints imposed by LCY stewardship over access to political power, there were no formal rules restricting political advancement along ethnic lines. But at the informal level, because of the quota system in place, advancement based on merit was curtailed. This produced two types of problems. One, the smaller ethnic groups faced the barrier of being able to advance only up to a certain level. Since the top posts were carefully apportioned by ethnicity and/or republic, a Serb or a Croat would be guaranteed the possibility of access to high posts. But the less numerous ethnic groups, especially those without an eponymous republic (e.g., the Albanians) and even the non-ethnically-defined "Yugoslavs," faced bigger hurdles. Their rise to the top posts could only be a result of political tradeoffs between the Serbs and Croats. In practice, members of ethnic groups without any administrative region faced a low ceiling on advancement in institutions of political power, even if they had substantial merit. Two, the other side of the coin was that members of the larger groups who were already overrepresented in political and security apparati (most of all the Serbs) had grievances of their own, based on the claim that they had to meet higher standards for promotion than those from other groups.

## Closure in the Economic Realm

The distribution of wealth in Yugoslavia was regionally unbalanced, with a relatively rich North and a poor South. Differences in the economic strategies and in the level of infrastructure of republics also followed a North-South pattern. While Slovenia, Croatia, Vojvodina, and parts of Serbia relied on attracting foreign investment and on building advanced production capacity, southern republics (Macedonia and Montenegro) depended on low-paying and labor-intensive activities and agriculture.

There is little information about the wealthiest individuals in Yugoslavia. In general, the wealthiest individuals appear to have been concentrated in the wealthiest republics, with Slovenes and Croats especially overrepresented. The tourist industry in those republics provided access to hard currency, the two republics were the most urban and industrialized, and a relatively much greater proportion of Slovenes and Croats had access to higher education than did the other ethnic groups in Yugoslavia.

At the general population level, net average pay was considerably higher in Slovenia—about $330 per month—than in Serbia, where workers on average took home about $260 monthly in net income. The figure was lowest in the autonomous province of Kosovo, averaging approximately $127 monthly. The regional disparity in wealth becomes even more obvious when comparing republic contributions to the gross social product (GSP),[6] shown in Table 3.9.

Data for 1989 indicate that the per-capita GSP of Slovenia was 2.5 times the per-capita GSP of Serbia and the autonomous provinces. Separate figures for Kosovo (unavailable) probably would indicate an even greater gap between the richest and the poorest parts of Yugoslavia.

Comparing the 1981 and 1989 figures points to a widening of the North-South divide during the 1980s. While the overall employment situation was not very favorable by the late 1980s (unemployment reached about 17 percent),[7] employment patterns varied substantially by republic, with Slovenia enjoying a relatively buoyant labor market and Kosovo suffering from serious unemployment.

------

[6]Yugoslav output was measured using gross social product (GSP), thought to be approximately 15 percent lower than GNP on average. Calculations for average take-home pay were roughly converted at 1990 exchange rates for purposes of comparison. See *Business International Forecasting Service: Yugoslavia,* Economist Intelligence Unit, March 1, 1991. Gross social (or material) product was equivalent to net material product (the normal CMEA measure) plus capital consumption. Like NMP, GSP excluded "nonproductive" services—such as education, health, defense, professional services, and public administration—and was thus not comparable to the Western concept of gross domestic product. National income represented the value of goods and productive services (including turnover taxes) relating to physical production, transport and distribution.

[7]*Country Profile: Yugoslavia,* Economist Intelligence Unit, August 27, 1991.

Table 3.9

Per-Capita Gross Social Product (GSP)

| | 1981 | | | 1989 | | |
| --- | --- | --- | --- | --- | --- | --- |
| | GSP | Population (thousands) | Per-Capita GSP | GSP | Population (thousands) | Per-Capita GSP |
| Serbia, including Kosovo and Vojvodina | 148.42 | 9,279 | 15,995.8 | 144.7 | 9,815 | 14,741.9 |
| Croatia | 101.18 | 4,578 | 22,101.6 | 101.2 | 4,726 | 21,422.9 |
| Slovenia | 64.57 | 1,884 | 34,271.1 | 72.5 | 1,924 | 37,705.4 |
| Bosnia-Herzegovina | 48.43 | 4,116 | 11,765.1 | 49.8 | 4,795 | 10,391 |
| Macedonia | 23.23 | 1,914 | 12,136 | 22.7 | 2,193 | 10,360 |
| Montenegro | 7.87 | 583 | 13,506 | 7.9 | 664 | 12,006 |
| Total | 393.70 | 22,354 | 17,612.1 | 398.8 | 24,117 | 16,544.2 |

NOTE: Per-capita GSP is given in thousands of YuD (Yugoslav dinars).

SOURCES: *Country Profile: Yugoslavia,* Economist Intelligence Unit, August 27, 1991. GSP data obtained from EIU Yugoslavia Country Profile, 1991.

**Assessment of closure in the economic realm.** One can confidently posit that Slovenes and Croats were disproportionately represented in the upper end of the income distribution, while Albanians, Macedonians, and rural Serbs comprised the low end of the distribution. However, it is likely that—controlling for relevant variables such as employment and education—the major determinant of income was not one's ethnic background but rather one's home republic or province. While there is a correlation between ethnic background and home republic, it is unlikely that ethnicity per se was a major cause of income disparities. The local economic climate and development strategies pursued by the various republics and autonomous provinces were the significant factors.

Was there a mechanism that allowed for a change of the "snapshot" presented above? As in any communist country, the state controlled the industry and placed severe restrictions on the type and size of private economic activity. The Yugoslav modification of the communist model was based on the self-management principle: a decentralized economy based on independent enterprises, some of which were privately owned. At the same time, self-management still placed far-reaching limits on personal wealth accumulation, and it did not remove the inefficiencies built into communist economies. With only a few exceptions, Yugoslav enterprises were not internationally competitive. Private economic activity was largely limited to small industry, agriculture, and services.

The federal government attempted to mitigate the economic differences between the North and South through a system of federal transfers of funds from northern to less efficient southern firms, informal subsidies, and the burden-sharing quotas established by the Federal Fund. These practices had secondary ethnic ramifications, for they meant an outflow of resources from the North to the South.[8]  As Table 3.10 shows, Slovenia's share of the burden

Table 3.10

**Percentage of Contributions and Disbursements from the Federal Fund**

| Contributions | Croatia | Slovenia | Serbia | Vojvodina |
|---|---|---|---|---|
| 1971–1975 | 34.52 | 22.62 | 29.53 | 13.33 |
| 1976–1990 | 34.27 | 22.02 | 30.28 | 13.23 |
| 1981–1985 | 33.60 | 20.73 | 32.55 | 13.12 |
| 1986–1988 | 31.80 | 25.63 | 31.20 | 11.37 |
| Disbursements | Bosnia-Herzegovina | Macedonia | Montenegro | Kosovo |
| 1971–1975 | 25.6 | 22.1 | 12.2 | 40.1 |
| 1976–1990 | 23.1 | 20.6 | 11.3 | 45.0 |
| 1981–1985 | 16.5 | 20.9 | 9.6 | 53.0 |
| 1986–1988 | 12.3 | 14.5 | 7.9 | 65.3 |

SOURCE:    Statistical Yearbook of Yugoslavia, 1989.    Adapted from Joseph T. Bombelles, "Federal Aid to the Less Developed Areas of Yugoslavia," *East European Politics and Societies,* Vol. 5, No. 3, Fall 1991.

---

[8]Joseph T. Bombelles, "Federal Aid to the Less Developed Areas of Yugoslavia," *East European Politics and Societies,* Vol. 5, No. 3, Fall 1991, p. 445. From a Yugoslav perspective, Federal Fund transfers from the more developed to the less developed parts of the state were simply regional development policies to benefit the country as a whole. But to many Slovenes and Croats, the federal transfers were little more than "forced loans."

increased steadily from 20.7 percent to 25.6 percent of the total Federal Fund over the 1981–1988 period; at the same time, contributions of Serbia, Croatia, and Vojvodina decreased slightly. The utility of the fund was limited; throughout the 1970s and 1980s, inequality between republics grew steadily despite the redistribution of resources at the federal level (in any given year, the fund would redistribute approximately 2 percent of gross social product). In ethnic terms, the wealth redistribution amounted to subsidies that penalized the Slovenes and Croats especially and rewarded the Albanians and Macedonians. Without the ethnic perspective, the wealth redistribution amounted to central reallocation of funds within the country, away from the wealthy to the poorer regions.

## Closure in the Social Realm

Status distinctions in Yugoslavia were clear and important. While the state took pains to ensure "equality" of ethnic representation, the very terminology used to describe and distinguish between the various ethnic groups only reinforced the different status accorded to them. Status at the national level stemmed from several sources, including the perceived "compatibility" of ethnic groups to the Yugoslav idea. A status stratification map for Yugoslavia in the late 1980s might have run along the lines presented in Table 3.11.

Status was based on a variety of distinctions, including Slav/non-Slav dichotomy, religion, level of development, and longevity as an "established nation." Yugoslav groups often perceived each other through an orientalist symbolic framework, which created a dichotomy between "civilized" and democratic groups on the one hand, and backward and authoritarian groups on the other. Such dichotomy was expressed in terms of "North versus South," "West versus East," or a combination of the two. Within the orientalist paradigm for determining national status, there were three major cleavages, determined by history of past rule, religion, and the "quality" of a people's nationalism.[9]

---

[9]Milica Bakic-Hayden and Robert M. Hayden, "Orientalist Variations on the Theme 'Balkans': Symbolic Geography in Recent Yugoslav Politics," *Slavic Review*, Vol. 51, No. 1, Spring 1992, p. 3; Milica Bakic-Hayden, "Nesting Orientalisms and Their Reversals in the Former Yugoslavia," *Slavic Review*, No. 54, No. 4, Winter 1995, pp. 917–931.

Table 3.11

Status Stratification

| Ethnic Group | Status |
|---|---|
| Slovenes | + + |
| Croats | + + |
| Serbs | + |
| Montenegrins | + |
| Macedonians | – |
| Muslims | – – |
| Hungarians | – – |
| Albanians | – – |

+ + = high status; – – = low status
(compiled on the basis of data
presented earlier).

A common perception among Slovenes and Croats was that their history of Hapsburg rule reinforced their higher level of political and economic attainment. Conversely, they attributed the poor economic and political performance of the "Balkan" republics to centuries of Ottoman rule. Many Slovenes and Croats also believed that their adherence to Catholicism kept their cultures firmly within the civilizational boundaries of "Europe," while Orthodoxy and Islam consigned their believers (Serbs, Muslims, Albanians, Macedonians, Montenegrins) to the "Byzantine" culture of the Balkans.[10] Moreover, nationalists in the northern republics often perceived their political agenda as being more sophisticated and "civilized" than those of nationalists in other republics. This was a view that distinguished

---

[10]The close correlation between religion, intensity of ethnic attachments, and rural status produced the most militant nationalists among farmers, whose traditional lifestyle and beliefs seemingly had not kept in line with the shifts among the urban population of the modernizing Yugoslav state. Empirical studies show a clear relationship between intensity of religious beliefs and ethnic intolerance in Yugoslavia. Randy Hodson, Dusko Sekulic, and Garth Massey, "National Tolerance in the Former Yugoslavia," *American Journal of Sociology,* Vol. 99, No. 6, May 1994, pp. 1534–1558. For a lengthier elaboration on the role of religious attachments in Yugoslavia, see Gerard E. Powers, "Religion, Conflict, and Prospects for Reconciliation in Bosnia, Croatia and Yugoslavia," *Journal of International Affairs,* Vol. 50, No. 1, Summer 1996, pp. 221–252.

between the supposedly "good" Western variant of nationalism in Yugoslavia and the "bad" eastern ones.[11]

The whole edifice of the orientalist set of prejudices and biases complemented the inherent status differences based on the structure of Yugoslavia. The very name of the state—translated as "South Slav land"—established a reference point for a sense of belonging and identity for people in Yugoslavia, and it reinforced a hierarchy of ethnic groups. The main south Slavic ethnic groups (Serbs, Croats, Slovenes, Montenegrins) were at the highest level. Each was a constituent "nation" of the first Yugoslavia that emerged after World War I.[12] National myths traced the continuity of each "nation" with previous states much further back in history  Although they were south Slavs, the Macedonians, as an Orthodox and a relatively new "nation" (post–World War II creation and previously considered to be either Bulgarians or Serbs), were lower than the other groups in status hierarchy.[13] Muslims were south Slavs too, but their non-Christian background and recognition in official nomenclature as a distinct "nation" only in 1968 placed them even lower in the hierarchy of Yugoslavia's ethnic groups. The Muslims' origins as Slavic converts to Islam during the Ottoman rule and their role in helping govern the territories for the Ottoman empire gave them a negative image of erstwhile "traitors" (a perception especially strong among Serbs, whose pejorative name for the Muslims was "Turks"). In any event, the less developed and more "Balkan" an ethnic group (Orthodox or Muslim, further south), the lower status it was accorded. By definition, neither Albanians nor Hungarians (as well as many smaller ethnic groups) were south Slavs, and their identification with a state called "South Slav land" rather than a neighboring state of co-ethnics was suspect in the eyes of the main Slavic groups.

All of these distinctions were informal and vehemently denied by officials, but they were implicit in the laws of the state. Formally,

---

[11]Hayden and Hayden, "Orientalist Variations on the Theme 'Balkans,'" pp. 5–12.

[12]The first Yugoslav state was called the "Kingdom of Serbs, Croats, and Slovenes." The Montenegrins were considered little more than "mountain Serbs" and were included in the name of the country in that fashion.

[13]Pedro Ramet, "Religion and Nationalism in Yugoslavia," in Pedro Ramet (ed.), *Religion and Nationalism in Soviet and East European Politics*, Durham, NC: Duke University Press, 1989, p. 300.

specific provisions in the Yugoslav constitution prohibited any status classification based on nationality. But to implement this ideal, an elaborate official vocabulary to describe ethnicity only reinforced the status differences. For example, *narod* (nation) was used exclusively in reference to the Slavic ethnic groups having only Yugoslavia as their constituent state: Slovenes, Croats, Muslims, Serbs, Macedonians, and Montenegrins. *Narodnost* (national minority) included those groups with a "mother state" outside the Yugoslav borders, such as Albanians and Hungarians (and dozens of numerically smaller groups, such as Turks or Italians) as well as those that had neither a state nor a Yugoslav republic (the Roma and Ruthenians were the most numerous of these).

The association of Serbs with the establishment of Yugoslavia and the "Yugoslav idea" (based on historical myths) led to the view held by many Serbs that their group had a right to leadership in Yugoslavia. Indeed, such views contributed to the initial development of a chasm in interwar Yugoslavia between Croats and Serbs. Much of the post–World War II history of Yugoslavia revolved around setting up sufficient checks and balances to prevent Serbs from assuming a formal and full leadership role in the country. But as the single most numerous ethnic group in Yugoslavia and with the same city serving as the capital of both Yugoslavia and Serbia, the view of Serbia and Yugoslavia as nearly synonymous was ever-present. This view was pernicious to the other ethnic groups, for it had the effect of elevating the status of Serbs as the primary ethnic group in Yugoslavia with a "natural" right to rule.

What kind of implications did the status distribution have? The decision by the drafters of the 1974 constitution to define Serbo-Croatian, Slovenian, and Macedonian as official languages was based on the status distinction between the *narod* and *narodnost,* as only those groups recognized as Yugoslav nations were entitled to have their language adopted for official use (though the constitution also guaranteed other ethnic groups the right to use their own language and alphabet). Status-based language restrictions were also found in the JNA, which did not implement constitutional provisions for language diversity and adopted Serbo-Croatian (with Latin script) as its only language of command, administration, and communication with outside parties.[14] Especially for non-Slavic groups, the language hurdle was a significant obstacle to overcome in order to function in the Yugoslav society.

Informally, a whole range of biases and perceptions arose around the status hierarchy, with the non-Slavic and Muslim ethnic groups perceived as the most backward and "foreign" elements, and even the Orthodox Slavic groups (Montenegrins, rural Serbs, or Macedonians) were seen as less sophisticated and talented than the Croats and Slovenes. The biases tended to be self-reinforcing and undoubtedly affected myriad everyday decisions.

Was there a mechanism to change the status stratification map? Status was tied closely to ethnicity, and there was very little potential for movement between ethnic groups. One's ethnicity was determined by birth and it was seen in ascriptive terms as something inherent. One's ethnic background was generally recognized through first and last names, and sometimes through distinguishing dress and/or physical characteristics. The political system implicitly used an understanding of ethnicity as something predetermined. Only generational-type movement between ethnic groups (intermarriage and consequent ethnic identification of offspring with the ethnic identity of one parent) was possible. The political system created an elaborate way of managing the ethnic relations, based on quotas, but it never transcended the problem of ethnicity being treated as a fundamental building block and a given. The quota system sharpened the importance of ethnicity and the status differences and biases based on ethnic distinctions. The ethnic quotas only illustrated the basic collectivist outlook upheld by the state; in essence, Yugoslavia was a federation of ethnic groups (rather than individuals associated on an administrative-territorial basis), held together and legitimated by a modified communist system.

The potential alternative to ethnicity as a defining building block came in 1971 in the form of a census category of "Yugoslav."[15] Five

---

[14]Anton Bebler, "The Military and the Yugoslav Crisis," *Südost-Europa,* No. 3, Vol. 40, 1991, pp. 127–144.

[15]Survey data (from 1985 and 1989) show interesting reasons for why some citizens of Yugoslavia chose to reidentify ethnically as "Yugoslav." There appear to have been four main routes: (1) young urban residents, (2) those whose parents came from

percent of Yugoslav citizens registered as "Yugoslav" in the 1981 census (this category dropped to 3 percent in the more ethnically tense conditions at the time of the 1991 census) but the category was not treated seriously in the quota-like system of apportionment of positions. Thus, in practice, the salience of ethnicity—and the attendant status differences based on ethnicity—was upheld by state policy.

The main obstacle to change in status distribution was the overall context of extremely limited mobility (only generational change) among ethnic groups. Other than that, no group was specifically targeted by state policy to inhibit further its mobility.

## Overall Assessment of Closure

Based on the information presented above, Table 3.12 summarizes the degree of closure (in an overall sense as well as in the political, economic, and social realms) experienced by Yugoslavia's main ethnic groups at the end of 1989. To reiterate, closure in Weberian terms refers to the "process of subordination whereby one group monopolizes advantages by closing off opportunities to another group." In the table, a group experiencing a "low" degree of closure has the most opportunities open to it. A group experiencing a "high" degree of closure has opportunities largely closed off.

As noted earlier, closure patterns in the political realm were related to the overrepresentation of Serbs and Montenegrins in the LCY and the armed forces relative to other groups such as Slovenes and Croats, as opposed to any ethnically specific policies to keep some groups out of power. The closure pattern in the economic realm was tied to the north-south development divide that favored the northern republics of Slovenia and Croatia. At the social level, the rigid status

different ethnic backgrounds, (3) Communist party members, and (4) persons from ethnic minorities. The third route is important in its indication of similarity of identity between communist and Yugoslav, but the fourth route is most telling, because of the seeming shame and sense of inferiority aroused by being a member of one of the small *narodnost* ethnic groups. But perhaps the most revealing evidence of the attitude toward "Yugoslav" identity was the census itself: the "Yugoslav" choice was subscripted with the explanation "having no identifiable nationality." Dusko Sekulic, Garth Massey, and Randy Hodson, "Who Were the Yugoslavs? Failed Sources of a Common Identity in the Former Yugoslavia," *American Sociological Review*, Vol. 59, February 1994, pp. 83–97.

Table 3.12

Patterns of Closure by Ethnicity in Yugoslavia

|  | Political | Economic | Social | Overall |
|---|---|---|---|---|
| Serbs | Low | Moderate | Moderate/Low | Moderate/Low |
| Montenegrins | Low | Moderate | Moderate/Low | Moderate/Low |
| Croats | Moderate | Low | Low | Low |
| Slovenes | Moderate | Low | Low | Low |
| Muslims | Moderate | Moderate | High | Moderate |
| Macedonians | Moderate | Moderate | Moderate | Moderate |
| Albanians | Moderate | High | High | High |
| Hungarians | Moderate | Low | High | Moderate |

differences were tied to the determinants of status (Slav/non-Slav background, relationship to the Yugoslav idea, and religion).

The "founding nations" of Yugoslavia—Serbs/Montenegrins, Croats, Slovenes—were the most privileged in the political, economic, and social realms. Serbs and Montenegrins faced some economic and social closure. But there is much to be said for the argument that many Serbs were predisposed to believe they faced much greater social closure than they actually did because of Serb mythologies (self-perception as an oppressed people, along with a martyr complex).[16] In fact, Serbs and Montenegrins were clearly in a dominant position in the most important realm—the political. Slovenes and Croats faced some closure in the political realm because of past events (purges in Croatia) and largely self-generated group-level disincentives to participate actively in Yugoslav LCY-oriented institutions.

The Albanians show a clear and consistent pattern of facing a high degree of closure; when compared to the Serbs and Montenegrins, the Albanians are the only major group that is worse off in each of the

[16]Marko Zivkovic, "Stories Serbs Tell Themselves: Discourses on Identity and Destiny in Serbia Since the Mid-1980s," *Problems of Post-Communism*, Vol. 44, No. 4, July–August 1997, pp. 22–29; Wolfgang Hoepken, "War, Memory, and Education in a Fragmented Society: The Case of Yugoslavia," *East European Politics and Societies*, Vol. 13, No. 1, Winter 1999, pp. 190–227.

three realms. Muslims, Macedonians, and Hungarians also faced some closure, but not to the same extent as the Albanians. Because of limited potential for change in the stratification patterns (in all three realms), the closure pattern was rigid.

A final ranking of groups along the lines of privileged to dominated—in relative terms—is seen in Table 3.13.

### Table 3.13

**Ranking of Ethnic Groups in Yugoslavia**

| | |
|---|---|
| Privileged | Slovenes |
| | Croats |
| | Serbs |
| | Montenegrins |
| | Macedonians |
| | Hungarians |
| ↓ | Muslims |
| Dominated | Albanians |

The specific placement of ethnic groups on the privileged-dominated scale is not evenly spaced. In other words, the range of difference between the top four groups was small relative to the wide gap separating the Muslims and especially the Albanians from the privileged groups.

Based only on relative deprivation, the Albanians seem to have had the biggest grievances and the most reasons for seeking to change the status quo. The Serbs were relatively privileged, though not as much as the Slovenes and Croats. The complicating factor was that an Albanian attempt to change the status quo would have to come at the expense of the Serbs (since the Albanians inhabited a part of Serbia and lived primarily in proximity to Serbs). On the other hand, as the main "founding nation" of Yugoslavia, the Serbs were relatively deprived in comparison to the Slovenes and Croats (primarily for reasons of self-perception and due to differences in levels of development). Further usurpation of their power and standing (by the Albanians) would have placed the Serbs at an even greater disadvantage vis-à-vis the most privileged groups. In this sense, the ethnic setup of Yugoslavia and the mechanisms in place to prevent ethnic tensions implied that the allocation of power and status followed a zero-sum game. Each group perceived its status relative to other groups, and could only improve its position by making other groups worse off. This, of course, sowed the seeds of instability and rivalry along ethnic lines.

## TRANSFORMING POTENTIAL STRIFE INTO LIKELY STRIFE

This section focuses on the process of Serb mobilization in the late 1980s by addressing the five aspects of mobilization: incipient changes, galvanizing "tipping" events, leadership, resources and organization, and the foreign element. Since this chapter is a retrospective look at a mobilization that already occurred, there is no ambiguity about which group was mobilized. However, an analyst looking at Yugoslavia in the late 1980s would have been wise to scrutinize all of the major groups in Yugoslavia from the perspective of what kind of events might lead each one to mobilize for political ends.

### Incipient Changes

There were at least four influential changes that, over time, shifted the demographic, economic, and political balance of power in 1980s Yugoslavia.[17]

**Shift in federal-republican power balance.**  The decentralization process in Yugoslavia had proceeded in fits and starts since the mid-1960s. The 1974 constitution was simply an expression of a long-term trend of decentralization and the shift of real power centers toward the republics. Over the course of two decades (mid-1960s to mid-1980s), Yugoslavia moved from a strong federation to a loose confederation. As long as Tito was alive, the decentralization and power shift was more in form than in substance, and Tito's authority

---

[17]The factors described here were interpreted by many ethnic Serbs as part of a conspiracy aiming to "weaken Serbia." Although analysts generally do not subscribe to the view that there was any conscious ethnically based anti-Serb intent behind them, the four trends did weaken the Serb position in Yugoslavia. Mojmir Krizan, "New Serbian Nationalism and the Third Balkan War," *Studies in East European Thought*, Vol. 46, 1994, pp. 47–68.

and—when necessary—use of repression ensured a meaningful federal policy. But with Tito's death in 1980, it became clear that the institutional arrangements he set up could not function without him. The rotating presidency and an extensive system of checks and balances led to a paralysis, a weakening of federal authority, and, eventually, an absence of any meaningful federal policy. Consequently, the republics took on a greater role in formulating and implementing policies that used to be the domain of the federal structures (such as foreign trade and foreign exchange policies). In other words, the republics truly implemented the far-reaching decentralization promulgated in the 1974 constitution. By the mid-1980s, the only federal structures with substantial influence were the LCY (and even the party was split up among republican lines), the secret police, the federal army (JNA), and—to a lesser extent—the Federal Executive Council's economic secretariats. The diminished role of federal structures implied a diminished Serbian role within Yugoslavia. It also meant that ethnic Serbs outside of Serbia proper were increasingly exposed to policies formulated by republican-level political structures controlled by other ethnic groups.

**Erosion of Serbian political-administrative ethnic unity.** Coupled with the changing federal-republican power balance, there was an increasing tendency toward the dispersion of ethnic Serbs among a variety of subfederal administrative units and the consequent dissipation of Serb power (in terms of the Serb ethnic group). The trend dated back to the early post–World War II period, with the drawing of republican borders that made Serbs the only ethnic group dispersed in several republics. The new interrepublican borders were even less favorable to Serbs; the war years witnessed substantial demographic changes, but the redrawing was also an attempt to diffuse Serb demographic power in the federation (to allay some of the fears of non-Serbs about Serb political power). The Serb dispersion caused by the new administrative borders was accompanied by the creation of a new "nation," the Macedonians, with their own republic, even though most Serbs previously considered Macedonians to be "southern Serbs." In addition, Montenegro attained the status of a republic, even though many Serbs considered Montenegrins to be "mountain Serbs." When Bosnia-Herzegovina reached republican status, another substantial portion of ethnic Serbs fell under the administration of a republic other than Serbia proper. Moreover, the elevation of Muslims to a status of an official "nation" in the 1960s and their increased influence in Bosnia-Herzegovina's power structures diminished the Serbs' position within that republic. Finally, the 1974 constitution elevated two regions in Serbia—Kosovo and Vojvodina—to the status of autonomous provinces and gave them political powers approaching in many ways those of republics. The administrative change in 1974 amounted to a de facto loss of sovereignty by Serbia over 36 percent of its territory. Of all the republics, only Serbia was affected in such a fashion (in other words, all the autonomous provinces in Yugoslavia had been carved out of Serbia).

**Demographic shift in Kosovo.** The different population growth rates among Serbs and Albanians in Kosovo (much higher among Albanians), combined with the province's autonomy and the quota-like system of ethnic representation, amounted to a long-term displacement of Serbs from positions of political influence in Kosovo. The demographic shift was accompanied by gradual but increasing signs of greater political assertiveness among the ethnic Albanians, which culminated in the request to transform Kosovo from an autonomous province within Serbia to a full-fledged republic of Yugoslavia. Kosovo occupied a special place in Serb mythology, as the "cradle of Serbia" and a host of Serb nationalist and religious (Serb Orthodox) symbols continued to be associated with Kosovo, making Albanian demands anathema to many Serbs with a predominantly ethnic view of Yugoslavia.

**Systemic economic malaise.** After an initial encouraging start, the self-management system failed to deliver the economic benefits promised by Yugoslav economic planners and increasingly came to be seen as a dead end. The economic inefficiencies of the system were aggravated by the oil shocks in the 1970s and the debt problems in the 1970s and 1980s. The lack of a coordinated macroeconomic policy (leading to a spiraling inflation) and the widening economic differentiation between the northern and southern republics showed the rigidity of the system and highlighted the need for its fundamental overhaul. In the Yugoslav context, the economic problems had an ethnic dimension because of the north-south split in terms of development and the transfer of resources from the richer north to the poorer south.

In addition to the four incipient changes, there was one major unexpected event that upset the balance of power in Yugoslavia in the late 1980s.

**The fall of communism.**  Besides allowing open questioning of the systemic fundamentals that underpinned the country (and, thus, the principles holding the country together), the fall and delegitimization of communism was also interpreted in Yugoslavia as a defeat for the Serbs, since the Serbs were most numerous in the LCY and identified most closely with the communist Yugoslav ideology.  Moreover, the fall of communism was accompanied at once by serious discussions about changes in state borders (German unification) and the elaboration of plans for rapid systemic change (the Polish "shock therapy" plan).  The former probably spurred the northern republics toward greater independence, while the latter opened up for them the possibility of a true evolution toward a free market system.

## Tipping Events

Three main tipping events elicited and propelled Serb mobilization.

**Two public memoranda by Serbian intellectuals in support of Serbian nationalist causes.**  The first document was a petition entitled "Against the Persecutions of Serbs in Kosovo," signed in January 1986 by 212 prominent Serbian intellectuals (associated with the Serbian Academy of Arts and Sciences and the Serbian Writers Union).  The second document was a memorandum circulated in September 1986 that linked the fundamental problems of Yugoslavia with the alleged attempt to splinter and weaken the "Serbian people."  It was authored by 23 members of the Serbian Academy of Arts and Sciences and was originally intended to be the product of a year-long effort to analyze the problems facing Yugoslavia.  In effect, the memorandum was a call to arms for Serbs in defense of the "Serbian people."[18]  Both documents gave the militant and paranoid form of Serbian nationalism legitimacy through their open acceptance and

---

[18]For the development of these views and their rise to prominence among an influential portion of Serbian intellectuals, see Nicholas J. Miller, "The Nonconformists: Dobrica Cosic and Mica Popovic Envision Serbia," *Slavic Review,* Vol. 58, No. 3, Fall 1999, pp. 515–536.

promulgation of chauvinist ideas, supported by some of Serbia's best-known and respected intellectuals.

**Two Kosovo appearances by Milosevic.** The Serbian LCY leader, Slobodan Milosevic, made two crucial public appearances at Kosovo Polje (in the province of Kosovo), the first in April 1987 and the second in June 1989. The former established the cause of Serbian nationalism (and the alleged discrimination faced by Serbs in Kosovo) firmly in the Serbian political discourse. The latter demonstrated the appeal of the cause of Serbian nationalism to Serbs, as nearly 10 percent of all Serbs in Yugoslavia gathered in Kosovo for a show of strength. Moreover, the speech delivered by Milosevic in 1989 took place in front of the Yugoslav state President Drnovsek and Prime Minister Markovic, and it contained clear allusions to the use of force by the Serbs.[19] The rural Serbs of Kosovo had attempted to make their grievances known to the Belgrade authorities since the early 1980s,[20] but Milosevic's adoption of the cause was crucial in the mobilization process. The strong association between Kosovo Polje and Serbian mythology (site of a battle in 1389 between southern Slavs and Ottomans) gave Milosevic's appearances a powerful symbolic appeal.

**The "Yogurt Revolution" in Vojvodina in October 1988.** As a result of Serb mass demonstrations in the capital city of Vojvodina on October 5, 1988, the party leadership in the province was replaced with Serbian nationalists loyal to Milosevic. The demonstrators surrounded the Vojvodina assembly building and cowed the whole leadership into resigning. The example of a successful and quick ouster of a provincial leadership as a result of extralegal street-level pressure was quickly followed by a similar ouster of the leadership in Montenegro and then in Kosovo. The elite replacements with Milosevic supporters paved the way for changes to the Serbian constitution in November 1988 that, in effect, abolished the autonomy of Vojvodina and Kosovo in March 1989. The mass demonstrations in

---

[19]Misha Glenny, *The Fall of Yugoslavia,* New York: Penguin Books, 1992, pp. 34–35.

[20]Sabrina Petra Ramet, "Nationalism and the 'Idiocy' of the Countryside: The Case of Serbia," *Ethnic and Racial Studies,* Vol. 19, No. 1, January 1996, p. 77; and Woodward, p. 88.

Vojvodina symbolically launched the drive to make "Serbia whole again."

## Leadership

The undisputed leader of Serbian mobilization was Slobodan Milosevic, an able bureaucrat but little-known until 1984. There is little in Milosevic's background to suggest any strong ambition for power, though his intelligence and effectiveness in organization were evident in his early years.[21]

Milosevic came from a poor family (of Montenegrin parents) in eastern Serbia; he joined the communist party while still in high school and, being a very good student, went on to law school in Belgrade. An effective organizer, Milosevic had a number of party posts in the Belgrade party organization early in his career before moving on to managerial positions in the state economy (including a stint as president of a large bank in Belgrade). He began to have an impact on Serbian politics after being promoted to head the Belgrade City LCY committee in 1984. Through the influence of mentors and by building a solid base of allies, Milosevic became the head of the Serbian party organization in 1986. He showed Machiavellian ruthlessness (deposing his erstwhile mentor and patron, Ivan Stambolic, in late 1987), good political instincts, and an extreme adaptability of principles (evident in his switching back and forth from orthodox communism to populist antibureaucratic sloganeering to Serbian nationalism). He reached the apex of his power in November 1989, when he was confirmed handily as the president of Serbia in a referendum-style election. Although ethnic Albanians boycotted the election, his widespread support among Serbs was genuine and undeniable.

The crucial factor behind Milosevic's success consisted in his ability to appeal to different powerful constituencies at once. He promised communist orthodoxy to LCY conservatives, free market reforms to liberals, the safeguarding of their rights to Kosovo Serbs, and the new

---

[21]For more background on Milosevic, see Veljko Vujacic, "Serbian Nationalism, Slobodan Milosevic and the Origins of the Yugoslav War," *The Harriman Review*, Vol. 8, No. 4, December 1995, pp. 25–34, and Aleksa Djilas, "A Profile of Slobodan Milosevic," *Foreign Affairs*, Vol. 72, No. 3, Summer 1993.

rise of Serbia to Serb nationalists. He even won the initial support of Croat and Slovene republican leaders by promoting himself as an antisecessionist, pro-Yugoslav federation figure intent on clearing the path for reforms.

By all accounts, Milosevic seems to have been genuinely surprised by his meteoric rise as charismatic leader (in the Weberian sense) after his speech in Kosovo in April 1987. However, he quickly became single-minded in his devotion to the cause of Serbian nationalism and channeled his considerable organizational and intellectual skills to furthering that cause and especially to establishing himself as the unrivaled leader of the new Serbian nationalist movement.

## Resources and Organization

The mobilized Serb movement was able to control and use the state machinery in the republic for its own ends. In effect, control of the Serbian state and party apparatus meant that the usual mechanisms for extraction of resources in the republic (i.e., taxes) could be harnessed to support the mobilization process and/or deny resources to those opposing it. Although Serbia had a lower per-capita level of wealth relative to Slovenia and Croatia, its large size gave it a substantial resource base at an absolute level. And, in a relative sense, no potential opponents of the mobilization within Serbia stood a chance in terms of access to resources.

Milosevic's position as the head of the Serbian party organization was crucial, as no other organization in Serbia came close to the Serbian LCY in terms of its knowledge of the resource base, the machinery to extract the resources, and the personnel to use for such ends. In addition, Milosevic was able to use his position to influence the executive and legislative branches of the republic. With access to these institutions came the ability to push legislation and other measures to extract resources (monetary, status, and positions of power) and distribute them to groups of crucial importance to the mobilization. Throughout the 1986–1989 period, Milosevic employed his bureaucratic power to support an elaborate pro-Serbian nationalist movement—aptly named the "antibureaucratic revolution."

One example of the use of state extractive machinery to support the mobilization process was the creation of "demo" networks—groups of disgruntled young men (often unemployed) paid to participate in nationalist gatherings throughout Serbia, Kosovo, and Vojvodina. Support for demo networks was obtained from commercial enterprises, which were either encouraged or forced to contribute to the costs of organizing rallies.[22]

The control and use of LCY and state machinery in Serbia for the mobilization provided a highly developed bureaucracy to support the process. The bureaucracy included party and state administrative cells throughout Serbia. Because of Milosevic's top party position, swift purges of dissenters ensured compliance and effectiveness within the bureaucracy. The control of the Serbian LCY also led to the quick harnessing of the central media (such as the mass-circulation *Politika* newspaper and Belgrade radio and television) for the purposes of mobilization.

In a step to build an even more loyal organization to support the mobilization, Milosevic created in 1988 the "Committee for the Defense of Kosovo Serbs and Montenegrins." The committee, with branches throughout Serbia and Montenegro, overlapped the LCY but was distinct from it. At the local level, branches of the committee acted as the vanguard of the mobilization—often using the contacts and resources of the LCY or the republic. The committees organized public demonstrations or shows of support, called "meetings of truth" or "solidarity" with Serbs from Kosovo. In 1988–1989, the committees organized more than sixty such meetings in Serbia, Montenegro, Kosovo, Vojvodina, Bosnia-Herzegovina, and Croatia, with a cumulative attendance in excess of 3.5 million people. Such meetings precipitated or contributed to the collapse of the Vojvodina, Kosovo, and Montenegro leaderships.

In addition, the Serbian Orthodox Church was co-opted to support the process of mobilization. The church hierarchy supported the mobilization, in part due to its inherent anti-Catholic (anti-Croat) and anti-Islamic (anti-Albanian and anti-Muslim) positions. Specific perks that Milosevic promised the church, such as allowances to build more shrines, probably also played a part in securing the support of religious leaders. The church's network of parishes provided

---

[22]Glenny, p. 34.

another organizational link to support the mobilization, allowing the process to reach elements that might otherwise have been hostile toward the communist and/or state apparatus, and, more important, it provided the organizational vehicle to reach ethnic Serbs living outside of Serbia (Croatia and Bosnia-Herzegovina).

In short, the Serbian ethnic mobilization between 1987 and 1989 was the best-organized political movement in Yugoslavia in the late 1980s. By taking over the Serbian LCY and state apparati, it obtained a complete hold on Serbia's political resources and institutions. Its reliance on both bureaucratic authority and traditional Serb symbolism allowed the movement to increase its ability to command resources, widen its support base, and avail itself of a pool of zealous Serb activists. By the end of 1989, the leader of the movement, Milosevic, was president of a stronger, centralized Serbia and could count on a friendly Montenegro.

## Foreign Element

There is little evidence to support the claim that ethnic Serbs living outside Yugoslavia were important in the mobilization of Serbs living in Yugoslavia. Nor is there any evidence that the Serbian ethnic mobilization was supported in any significant fashion from abroad.

## Overall Assessment of Mobilization

The long-term centrifugal trends in Yugoslavia (accelerating after Tito's death), the sudden demise of communism that delegitimized the political foundation of Yugoslavia, and, most important, the co-opting of the Serbian nationalist cause by a skillful and opportunistic communist leader combined to produce a powerful movement. The control and manipulation of the communist and state apparatus within Serbia to support the mobilization gave the movement access to tremendous organizational and resource bases, making it virtually unstoppable within Serbia and a powerful force within Yugoslavia as a whole.

## ASSESSING THE STATE

### Accommodative Capability

How inclusive and responsive were the Yugoslav political structures to popular will?  The 1974 constitution established a set of ethnically inclusive federal institutions, with a multitude of guarantees of equitable and ethnically proportional access to power.  However, the importance of LCY membership established indirect constraints on access to power because most ethnic groups were less attracted to LCY than were Serbs and Montenegrins.  The high level of Serb involvement in the LCY organization meant that this group was over-represented in the pool of party activists and potential candidates for office, state bureaucracy, and the armed forces.  The system also remained essentially authoritarian, with LCY the only legitimate political party, a heavy reliance by state institutions on the LCY for staffing, and a lack of free elections.  Thus, the system was only partially accountable and was inclusive only to the extent that it was open to those who accepted a communist federal state.

What kind of potential for change in political structures existed in Yugoslavia in the late 1980s?  The State Presidency, the LCY Central Committee, and the Federal Executive Council often acted as power brokers in settling disputes among regional interests.  The post-Tito federal system was unwieldy and prone to paralysis and gridlock. For example, federal constitutional amendments needed to be ratified by the federal parliament and by the eight republican and provincial units before entering into force.  The consensus rule meant that republics had veto power over any initiative that threatened their interests.  The need to take into account so many administrative units made the process of institutional reform on any fundamental issue next to impossible.  Finally, the whole edifice of communist party control (or at least oversight) of all significant political structures was rigid and not amenable to meaningful reform.  The potential for fundamental change in central political structures was low, since the system was designed to prevent any strong central rule.  The possibilities for change and reform at the republican or provincial level had their bounds set by the basic requirement of LCY oversight of the process.

In short, the Yugoslav system aimed at a high level of ethnic inclusiveness within the bounds of some important filters, especially the LCY as the vehicle to political access and power. The system was inflexible in that it protected LCY's monopoly role and made the federal bodies so weak as to prevent any central-level responsiveness, channeling responsiveness to republican- or provincial-level bodies.

Overall, the assessment is difficult. The decentralized Yugoslav system was far more responsive and accountable than orthodox Soviet-style communist systems, but it remained less responsive than a system with the access to power open to a wide range of views and interests. Ultimately, the best that can be said is that the Yugoslav system was the most responsive among communist systems. Once communism became delegitimized, the frame of reference changed (when it was compared to other communist systems, the Yugoslav system looked relatively responsive, but compared to prevailing liberal systems in Europe it did not fare so well).

What were the prevailing norms of governance? Since the 1974 constitution, the Yugoslav political process relied on consensus for most of its decisionmaking. This applied at both the federal and republican levels. The system tolerated considerable range of differences at the republican and provincial levels, but it guarded the LCY monopoly on power and did not hesitate to use force to prevent any challenges to that monopoly. The norms of tolerance extended only to the acceptance of dissent and a range of views within the structure of a federal communist Yugoslavia. No groups were purposely excluded from governing the country, but they had to go through the filter of LCY membership. The Yugoslav authorities expended considerable effort trying to make the LCY more attractive to members of some of the less represented ethnic groups, such as the Albanians. But the association of Serbs with the LCY and the existence of group-level antipathies between Albanians and Serbs as well as some group-level distrust between Croats and Serbs limited the LCY's attractiveness to many non-Serb or Montenegrin Yugoslav citizens. Polling data on intergroup antipathies were proscribed in Yugoslavia, but secondary evidence (clearest in the delineation of social status) suggests the persistence of residual distrust between groups and, in some cases, even hostility. Finally, collectivist norms (ethnically based) were clearly superior to individual rights, underpinned the whole structure of governance, and conditioned individual responses. In other words, the ethnically based manner of classifying individuals was a pillar of Yugoslavia, upheld by all Yugoslav institutions.

What was the level of cohesion among the ruling elites? At the federal level, the Yugoslav officials were divided substantially in their outlooks and deadlocked on the direction of the future evolution of the state. The veto power awarded to all republics made consensus necessary for decisions, and the different policy outlooks of the republics made consensus next to impossible to achieve. To varying degrees, all republican governments had plans to reform Yugoslav institutions at the end of the 1980s. The blueprints varied from further decentralization (proposed by the Slovenes) to strengthening of the federal organs (proposed by the Serbs under Milosevic). Economically based evaluations of the costs and benefits of a changed federal structure (with the southern republics having the most to lose from further decentralization) also affected the outlooks. The growing economic malaise made consensus even more difficult, with many republics protesting against the austerity measures proposed by prime ministers Mikulic (who resigned when the Federal Assembly failed to adopt his 1989 budget) and Markovic. The lack of cohesion among the ruling elites precluded any change in the way conflict was mediated at the federal level and contributed to the gridlock.

In conclusion, the accommodative capability of the Yugoslav state was low, as it was principally a function of the accommodative predispositions of the republics, channeled through a federal body. No Yugoslav institution was truly above the republics, a situation that made the federal state almost powerless in its attempt to forge or to force consensus on crucial reform decisions.[23] The consensus-based approach to policymaking was suited for a period in which different interests maintained reconcilable political agendas. Such agendas—at least in theory—remained compatible as long as there was a common acceptance of communist ideology and a decentralized federal structure as the guiding principles in Yugoslav politics. Milosevic's ethnic mobilization of the Serbs and his attempt to recentralize Yugoslavia (with the Serbs playing a much greater role in such a state) subverted the state from within, paralyzing its already

---

[23]Woodward, pp. 84–85.

low ability to adapt and to change. A paradoxical situation ensued, in that the leaders of the Serbian ethnic mobilization against the state also attained important posts at the state and federal levels.

## Fiscal and Economic Capability

The fiscal health of Yugoslavia was precarious at best throughout the 1980s, with several macroeconomic indicators registering economic stagnation. Deficit spending reached alarming levels by the mid-1980s. Moreover, the decentralized Yugoslav state had little power to increase its revenues or implement radical economic reform, as the republics had control over many macroeconomic tools and trade instruments.

While the 1953–1981 period saw Yugoslav gross social product (GSP) average an annual growth of 6.7 percent, in the 1980s GSP growth slowed considerably, principally as a result of the 1979–1980 rise in oil prices and the increased burden of interest payments and principal repayment on accumulated debt. The economic crisis at the end of the 1980s caused a steady decline in GSP in 1987 and 1988, with a slight increase in 1989 (see Table 3.14).[24]

### Table 3.14

### Trends in Yugoslav Gross Social Product

|  | 1985 | 1986 | 1987 | 1988 | 1989 |
|---|---|---|---|---|---|
| Total |  |  |  |  |  |
| At current prices (YuD bn) | 1.1 | 2.2 | 4.9 | 14.9 | 221.4 |
| At 1972 prices (YuD bn) | 393.7 | 407.8 | 403.1 | 397.1 | 398.6 |
| Real change (%) | 0.5 | 3.6 | −1.2 | −1.5 | 0.4 |
| Per capita |  |  |  |  |  |
| At current prices (YuD) | 47.6 | 94.5 | 209.0 | 632.2 | 9,345.7 |
| At 1972 prices (YuD) | 17,029 | 17,525 | 17,212 | 16,848 | 16,826 |
| Real change (%) |  | 2.9 | −1.8 | −2.1 | −0.1 |

YuD = new Yugoslavian dinars.
SOURCES: Indeks; National Bank of Yugoslavia, adapted from the EIU Yugoslavia Country Profile, 1991.

---

[24]*Country Profile: Yugoslavia,* Economist Intelligence Unit, August 27, 1991.

Inflation continued to spiral upward, reaching 150 percent in 1987 and 1,950 percent in 1989. In 1989, the total Yugoslav debt was over US$20 billion. In per-capita terms, Yugoslavia had one of the highest foreign debt levels in Europe (see Table 3.15).

From 1987 onward, the Yugoslav federal government engaged in substantial deficit spending. In 1987, revenues contributed less than half to federal spending; the deficit situation remained serious but improved in 1988 and 1989. The 1980s also saw the rapid depreciation of the dinar, which eventually raised import prices and contributed to a slowdown in import growth and low rates of consumer and investment expenditure.[25]

By the late 1980s, the resource extraction potential of the federal government was limited at best. The slowdown in domestic consumption and economic reform decreased the federal revenue base. This problem was compounded by the fact that republics were entitled to the same taxable resources as the federal government (mainly turnover taxes and assessments by local and federal self-management communities) in addition to income and personal property taxes. The 1974 constitution virtually eliminated direct federal expenditures on investment—partly for this reason, in 1990 the federal government accounted for only 25 percent of total government

Table 3.15

Yugoslav Debt Data, 1984–1988 (in US$ million)

|  | 1984 | 1986 | 1988 |
| --- | --- | --- | --- |
| Gross national product | 44,274 | 64,664 | 49,782 |
| International reserves | 1,732 | 2,189 | 3,074 |
| External debt, excluding IMF | 17,691 | 19,414 | 20,373 |
| Principal repayments | 1,567 | 1,540 | 1,773 |
| Net flows | −120 | −886 | 52 |
| Interest payments | 2,338 | 1,777 | 1,401 |

SOURCE:  World Bank, World Debt Tables, 1989–1990.

---

[25]*International Economic Appraisal:  Yugoslavia,* Economist Intelligence Unit, February 29, 1988.

spending in Yugoslavia. The impotence of federal structures on fiscal matters was aggravated by the inability to set a strong monetary policy. In fact, reforms in the 1970s limited the control of the National Bank of Yugoslavia over commercial banks and made it almost powerless to carry out national monetary policy.[26]

In all, by 1989, the Yugoslav state was not backed by a readily identifiable ruling elite—the six republics formed a loose core constituency that supported the federal structure. The fact that the wealthiest republics in Yugoslavia were unwilling to increase their financial contributions to the central authorities in Belgrade (given their own revenue problems resulting from the general economic contraction) made the cash-strapped federal government even weaker.

In conclusion, all indicators show that by the late 1980s, the fiscal and economic resources and capabilities available to the Yugoslav federal ruling bodies were extremely low.

## Coercive Capability

Yugoslavia in the 1980s had four major security institutions, each with a separate command and control arrangement:  The Yugoslav People's Army (or the JNA, the regular armed forces of Yugoslavia); the Territorial Defense Forces (TDF); forces controlled by the Secretariat for Internal Affairs; and Republican and provincial police forces.

The 1974 constitution named the Presidency as "supreme body in charge of administration and command of the armed forces." However, the 1974 constitution also limited the scope of JNA interventions in internal affairs. In addition, the JNA's mission to protect the constitutional order was under the authority of the federal presidency and therefore subject to a majority vote among the republic chiefs. Within the constitutional framework, the JNA would not have been able to turn legitimately against a republican government.

---

[26]Because the credit policies of commercial banks were relatively unchecked and because they were organized on a republic basis, the banks were powerful in maintaining the serious imbalance of investment and development among the regions of Yugoslavia.

Operational command of the armed forces rested with the minister of national defense.

The TDF were formed in 1968 as an integral part of the Total National Defense Doctrine (aimed at denying the use of territory to the enemy through a total war of all citizens, using guerrilla tactics if necessary).  The TDF were mobilizable forces set up to prevent control of territory by an invader, and they were characterized by a high degree of decentralization and independence.  While responsible to JNA commands, TDF units were funded by local LCY bodies and were commanded by local TDF officials.  This mixed command arrangement was a source of friction between the republics and the JNA, and led to a centralization drive by the JNA to consolidate its control over the TDF (since the republics were still required to provide infrastructure and logistical units operating on their territory).  The Kosovo TDF was dismantled after the Albanian riots in 1981.[27]

The Secretariat for Internal Affairs controlled the State Security Service (SDB) and the People's Militia.  The People's Militia, which numbered more than 15,000 troops, operated numerous armored vehicles equipped with machine guns, water cannons, smoke and tear gas launchers for crowd control and riot situations, armored personnel carriers, and helicopters.[28]  These internal security troops were well paid, heavily indoctrinated, experienced, and reliable.  They could be deployed in times of political unrest or disorder when the local police were expected to side with the populace against federal authorities.[29]

The substantial republican role in securing internal order was granted by the National Defense Act of 1974.  In addition to its militia (police) forces and intelligence agencies, each republic had its own Secretariat of Internal Affairs, which maintained control over special forces with specific riot-control capabilities.[30]  After the Serb takeover in Kosovo, large numbers of special units from Serbia's Sec-

---

[27]Bebler, p. 137.

[28]Isby, p. 395.

[29]The Secretariat for Internal Affairs also controlled 15,000 troops in border guard units and a coast guard (part of the border guard) comprising sixteen patrol boats in 1990.

[30]Isby, p. 397.

retariat for Internal Affairs were deployed in the province.  The Kosovo and Vojvodina militia and internal affairs forces came under the direct control of Serbian authorities after the takeover.  While the federal Secretariat for Internal Affairs maintained nominal control over republican counterparts, events in Serbia demonstrated that the latter enjoyed significant operational autonomy and, in fact, was more responsive to the republican than to the federal authorities.

Who served in the apparati of violence?  Serbs and Montenegrins were overrepresented in the officer corps of the armed forces, and especially in the ground forces.  At the conscript level, however, every JNA unit included soldiers of each of the main ethnic groups.  With the exception of the Serbs, conscripts usually were not trained or stationed in their home republics or provinces.  This practice in theory ensured troop loyalty during internal security actions by the army.[31]

Although composition data of the TDF and republican internal security forces are not available, it is safe to assume that membership of these units roughly reflected the ethnic composition of their home republic or province.  Given the general overrepresentation of Serbs in the federal bureaucracy, one would also expect the SDB and the People's Militia to have been predominantly staffed by ethnic Serbs.

What kinds of norms were there in place toward the use of violence domestically?  Until Milosevic's rise in the middle and late 1980s, the state had used force against a number of attempts at ethnic mobilization, whether the mobilization took place through LCY channels or outside of them.  It had a reputation for guarding jealously the monopoly role of the LCY in Yugoslav politics and dealing harshly with any opponents.

Internal deployments of security forces occurred in Kosovo throughout the 1980s.  Small-scale disorders were quelled by units from the federal and Serbian Secretariats for Internal Affairs, the People's Militia, the SDB, and the local militia.  The JNA became involved in

---

[31]This principle was used also in other communist countries, including the USSR. The concept was that, for example, Macedonian soldiers would likely have fewer reservations about using force to restore order among the population of Croatia than against their fellow Macedonians.

Kosovo when the riots escalated to general unrest in 1981. Under a declaration of national emergency, the army intervened to stop demonstrations by ethnic Albanians beyond the control of the People's Militia and local militia. Hundreds of citizens were injured, and some were killed during the JNA's suppression of the demonstrations. Up to one-fourth of the JNA's total manpower remained in Kosovo to maintain order throughout the 1980s.

More important, the armed forces maintained different norms toward domestic use (or threatened use) of force depending on the republic. The JNA showed willingness to intervene and uphold orthodox communist and pro-Yugoslav values in Slovenia in 1988–1989. On the other hand, the rise of Milosevic to power and the takeover of the Kosovo and Vojvodina governments was either applauded or not opposed by a wide segment of the officer corps.[32] JNA support for Milosevic seems to have been based on ideological affinity rather than on ethnic support for his nationalist program. However, Serb overrepresentation in the armed forces may have inhibited the possibility of armed intervention against Serbian ethnic mobilization.

Was the force suitable for domestic use? The apparati of violence had limited preparation to handle low-level domestic conflict. The federal military, committed to its Total National Defense doctrine, was a modern conventional force, lacking the capabilities or training to effectively handle internal conflict. The TDF had a defensive regional focus and were neither suitable for nor relied upon by federal authorities for internal security. The People's Militia constituted a well-armed and trained force able to quell even large riots. The special riot control units of republic-level internal security forces also could handle some domestic unrest. However, no security organ in Yugoslavia had a rapid-reaction force for the prevention of serious, large-scale uprisings and conflicts. The lack of military units earmarked for controlling serious disorder had negative repercussions in Kosovo. Ethnic Albanian rebellions that could not be handled by Internal Affairs forces were suppressed by heavy-handed JNA interventions.

---

[32]Laura Silber and Allan Little, *Yugoslavia: Death of a Nation*, New York: Penguin Books, 1997, pp. 58–69.

In conclusion, the coercive means of the Yugoslav state against internal opponents were substantial in terms of riot suppression and the number of potential apparati that could be used. But these apparati lacked the capabilities to tackle low-intensity conflict (security threats that were more serious than riots but stopped short of war). In addition, some key internal security organs, such as the republican secret police and militia, often disregarded the formal supervisory role of the federal Secretariat for Internal Affairs and answered to the republican authorities.

The most important limitation on the apparati of violence in late-1980s Yugoslavia were the constraints on their use against a constituent part of the Yugoslav state. Because deployment against a republic or province relied on consensus, it was likely that the given republic or province would veto the deployment. Against such objections, deployment of federal forces would amount to the breaking of specific provisions of the constitution. Complicating the matter, the TDF were regionally focused and provided readily available sources of military expertise and assets that could be exploited by republics. In case of a supraconstitutional deployment of the federal forces, the TDF could serve as the core of the republican armed forces, raising the specter of a civil war.

## STRATEGIC BARGAINING

All the assessments so far regarding the mobilization of ethnic Serbs and the capabilities of the Yugoslav federal state to deal with such a mobilization provide the points of reference for thinking about the interaction between them while using the categories and matrices of the framework presented in Chapter Two. This section categorizes the group and state types on the basis of their capacities. The matrices provide a way to think conceptually about their interaction.

### Measuring the Group's Capacities

Concerning the leadership of the mobilized Serbs, all of the observations compiled imply that Milosevic was confident and secure in his position by early 1990. His victory in a popular referendum in late 1989 consolidated his standing and gave him prestige that no other Yugoslav leader could approach. Milosevic took risks and gambled (usually with success) throughout his meteoric rise to power. He appealed to all constituencies when he needed to but did not hesitate to paint as adversaries any number of groups. Although seemingly surprised by the rapid and widespread support his nationalist rhetoric provoked among ethnic Serbs, he seized upon the issue with determination and single-mindedness. Thus, the assessment of leadership is "strong."

As for the resource support of the mobilized Serbs, all of the earlier observations imply that the coalition Milosevic orchestrated had good resource support, in both an absolute and relative sense. The mobilized ethnic Serbs had sufficient support to meet all near-term objectives (reaching all ethnic Serbs so as to include them in the mobilization process) as well as prospects of even greater support (from Montenegro and the ethnic Serbs of Bosnia-Herzegovina and Croatia) as the movement gained momentum. The available support was suited to the goal of recentralizing Yugoslavia under a more dominant Serbian leadership. Thus, the assessment of resource support is "good."

Regarding popular support for Serb mobilization, Milosevic and his allies had substantial support among ethnic Serbs. Group mobilization had proceeded to focus on the ethnic Serbs inhabiting Serbia, but the ethnic Serbs in other republics provided an expandable base of support. Moreover, sympathy or potential for coerced support for the mobilization existed among other ethnic groups (Montenegrins and Macedonians, respectively). Thus, the assessment of popular support is "broad."

Based on these assessments, the mobilized Serbs are a type A group. The capacities of such a group are as follows:

   Accommodative: high;

   Sustainment: high;

   Cohesiveness: high.

## Measuring the State's Capacities

Concerning the leadership of the state, all the observations compiled paint a picture of the Yugoslav federal structure having an extremely weak leadership capacity. Indeed, an independent federal leadership was difficult to identify because it was so constrained by the will of the republics. Many federal institutions were chaired by republican-level elites, more loyal to the republic than to the federal structures. The collective federal leadership was intimidated by Milosevic and conscious of his popular base of power. It was neither willing nor able to make decisions, let alone take risks. The institutionalized conflict-defusing consensus rule meant that any republic had veto power and could prevent actions hostile to its interests. Thus, the assessment of leadership is "weak."

As for the fiscal position of the state, the federal regime was in a precarious and extremely weak situation. The state had engaged in deep deficit spending. With an eroding tax base and limited control over raising revenue, the federal machinery was in no position to spend more nor reallocate any significant funds. Opposition from several republics to any recentralization of economic power as well as a heavy foreign debt burden closed off any options for increased revenue generation. Thus, the assessment of fiscal position is "weak."

Regarding the regime type of the state, this capacity is not as clearcut as the others. The fundamental problem is that the regime was inclusive once past the "filter" of the LCY. Elections were not fully competitive, since the LCY had a monopoly on power. At the same time, the LCY "filter" was little more than a pledge of acceptance of a unified Yugoslavia with a modified state socialist system. A broad range of opinions existed in the sphere of how reformist the state socialist system should be (spanning the ideological space from a regulated market economy to orthodox communist views emphasizing the need for greater state control). The 1974 constitution also allowed for a relatively high degree of grass-roots representation at the federal legislative level. The media remained under some constraints, primarily to prevent ethnic nationalists from having a mouthpiece for their views (until Milosevic subverted the system from within). Limits on executive power were so far-reaching that they virtually stripped away most of the executive's usual powers and gave them to the republics. There were limits to the norms of tolerance of dissent, primarily centering on the idea of accepting LCY role and prevention of ethnic sloganeering. However, it is also clear that wide-ranging regional differentiation and devolution of power to the republics was accepted, and individual (republic-based) determination of economic developmental paths was recognized. Thus, despite some exclusion of non-LCY forces in the federal political process, the institutional arrangements in Yugoslavia remained more inclusive than not. The state expended considerable efforts to ensure that political access was proportionally distributed among ethnic groups, and it attempted to include citizens of all ethnic backgrounds in the political process. Finally, the analysis needs to take into account the nature of the challenging group. Since the filter of LCY actually heightened Serb influence in the state, it would make little sense to code the state as exclusive when thinking of how it might deal with Serb ethnic mobilization. If the challenge to the state came from mobilized Croats or Slovenes, there would have been more justification to code the state as exclusive. Thus, the assessment of regime type is "inclusive."

Based on these assessments, the federal Yugoslav state is a type E state. The capacities of such a state are as follows:

Accommodative: high;

Sustainment: low;

Coercive: low.

## Outcome of Bargaining and Preferences for Violence

Based on the matrix showing the preferences of the mobilized group, a type A group has the following preferences toward a type E state: (1) negotiate, (2) exploit, and (3) intimidate. Based on the matrix showing the preferences of the state toward a mobilized group, a type E state has the following preferences toward a type A group: (1) negotiate, (2) exploit, and (3) surrender.

Comparing group and state preferences leads to the striking conclusion that the potential for violence in the dyadic encounter between a strong mobilized ethnic Serb group and the federal state of Yugoslavia was low. The preferred Serb strategy was "negotiate," with a hedging strategy of exploitation or perhaps even outright intimidation in order to achieve its goals. The preferred federal Yugoslav strategy was "negotiate," with a hedging strategy of exploitation or outright surrender. Quite simply, in real-world terms, the federal Yugoslav state did not have the capacity to resist a determined Serb effort—especially based on republican authority—to subvert the state.

The choices of strategy bear out the options available to the two sides. If the preferred peaceful negotiations to recentralize the federation under Serbian direction had failed, the Serbs had the option of being more forceful in the bargaining process. They had the resources to do so, though they would have preferred a cheaper—peaceful—takeover of the federal structures. On the other hand, the federal state could only hope that the peaceful bargaining would succeed. A more forceful bargaining posture, exploitation, was more risky, for it was essentially a bluffing strategy. Against a strong mobilized group like the ethnic Serbs under Milosevic, the strategy risked that the bluff might be called. The surrender option as a third choice only illustrates that the range of choices for federal Yugoslavia was between peaceful or more forceful bargaining. The option of the use of force was not really available to federal Yugoslavia in that dyadic encounter. The state probably would capitulate in the face of more determined Serb moves.

What is telling about the choice of strategies is that the Serbs dealt from a position of strength. They could up the ante and escalate their threats in the bargaining process and back them up if necessary. In response, federal Yugoslavia could not counter the Serbs. It was unable to deal with the challenge and was likely to back down if the Serbs increased the pressure.

## The Course of Events

The strategic choices outlined above approximate closely the course of events. Serb goals became abundantly clear in 1990: either the federation would be amended to assure the protection of ethnic Serbs throughout the state, or the federation would be dissolved, with republican boundaries altered to create one single enlarged Serbian state.[33] The Serb goals were neither palatable nor acceptable to the other republics (save the Serbian ally Montenegro). The relatively wealthy Slovenia and Croatia had the most to lose, and they

---

[33]Steven L. Burg, "Why Yugoslavia Fell Apart," *Current History*, November 1993, pp. 357–363.

moved toward independence. Since the Slovenes recognized that their republic was of little interest to Serbia, they moved the fastest. The Serbs threatened force to elicit compliance from the other republics on a number of occasions and came close (one vote short in the federal Presidency) to succeeding. The behavior illustrates the "exploit" secondary strategy. The federal structures could neither prevent the Serbs from using force outside the constitutional framework (which they did from March 1991 onward, in a turn to an "intimidation" strategy) nor force them to back down. Indeed, by mid-1991, the state effectively chose the "surrender" option when it fell apart. In fact, state collapse was the final outcome of the bargaining process.

To be sure, the administrative machinery within the republics continued to function, though the republican role now increased to that of independent state actors because of the collapse of the federal state. And the inability of the federal state to deter ethnic Serb mobilization implied that other republics had to resort to their own means to resist the gradual increase of Serbian influence within the federation. Thus, the very low likelihood of conflict between mobilized ethnic Serbs and the federal authorities actually increased the likelihood of unmediated conflict between the various republics, eventually tearing down the constitutional fabric and splitting federal institutions along ethnic lines.

The republics and provinces where ethnic Serb mobilization was successful (Serbia, Kosovo, Vojvodina, Montenegro) formed a successor entity, still named Yugoslavia. The other four republics became independent. Whereas the absence of any significant ethnic Serb population in Slovenia made that republic of little interest to Milosevic, Croatia and Bosnia-Herzegovina soon plunged into lengthy strife consisting of a mixture of civil war and overt and/or covert intervention by Serbia on behalf of the ethnic Serbs in these two republics. Macedonia had few ethnic Serbs and rated low in the Serb mobilization scheme; the preventive deployment of outside forces on Macedonian territory also may have helped to keep that republic out of the wars of Yugoslav succession.

Had an analyst, at the end of 1989, used a framework similar to the one presented here to examine the situation in Yugoslavia, the following intelligence needs would have become apparent:

- The interrelationship and coordination between the channels of ethnic Serb mobilization (Serbian communist party and administrative apparatus, Orthodox Church, and the "committees for the defense of Serbian people");

- The influence of Milosevic's supporters in the Yugoslav apparati of violence;

- Countermobilization strategies in republics outside of Serbia;

- The means of control by Milosevic over provincial and/or republican leaders allied to him or installed by him;

- The potential appeal of non-ethnically-driven postcommunist evolution in Serbia and Montenegro;

- The non-Serb republics' willingness to increase the strength and resources of the federation.

Greater attention to these topics might have allowed a better preparation for the breakup, or the formulation of policies that might have prevented a violent breakup.

## FINAL OBSERVATIONS

The framework presented here deals only with the dyad of the mobilized ethnic Serbian group and the federal Yugoslav state. The model was applied to analyze the initial steps of ethnic mobilization and rivalry that eventually led to the disintegration of Yugoslavia. The model is not suitable for examining the events immediately preceding the outbreak of the wars of succession in the final stages of Yugoslavia's unraveling (early 1991), as those conflicts have more to do with interstate wars in highly fluid conditions than with bargaining processes internal to one state. Deterrence theory may be more fruitful for modeling the strategic choices open to the various republics in Yugoslavia in late 1990 and early 1991. The case examined here is interesting in the sense that a strong leader appealing to ethnic attachments rose within the governing structure of Serbia and, consequently, used the political institutions in Serbia against the Yugoslav state.

Use of the model in reference to the breakup of Yugoslavia illustrates the following main points:

- Long-term trends of diminishing Serb influence within Yugoslavia and a zero-sum competition between the ethnic Albanians and Serbs in Kosovo propelled ethnic mobilization of Serbs in the middle and late 1980s;

- The hijacking of the Serbian communist party and republican apparatus by the leaders of the Serbian ethnic mobilization made the movement a virtually unstoppable force within Serbia and an extremely potent force within Yugoslavia as a whole;

- The federal Yugoslav state was too weak to deal with a determined challenge from its strongest constituent part;

- The preferred strategy of the mobilized ethnic Serbs vis-à-vis the federal state was peaceful renegotiation of the arrangements governing the federation, but the secondary and hedging strategies of the movement relied on force;

- The preferred strategy of the federal state vis-à-vis the mobilized ethnic Serbs was peaceful negotiation, since it lacked the means to back up more forceful strategies;

- The potential for violence in the dyadic encounter between a strong mobilized ethnic Serb group and the federal state of Yugoslavia was low;

- The very low likelihood of conflict between mobilized ethnic Serbs and the federal authorities actually increased the likelihood of unmediated conflict between the various republics, eventually tearing down the constitutional fabric and splitting federal institutions along ethnic lines.

In retrospect, the strategic choices identified for each side were followed closely as events unfolded in 1990–1991. In that sense, the model's accuracy was validated with respect to the breakup of Yugoslavia. The model captured well the strategic preferences of the two actors. And although the analysis undertaken here used January 1990 as the cutoff date for data gathering, the same analysis could have been undertaken in mid-1987, with a similar outcome. While 1987 was an early point in the mobilization of the ethnic Serbs, most of the crucial mobilization factors were already in place and should have appeared as important elements of an intelligence assessment at that time. In addition, the capacity of the federal state did not change substantially between 1987 and 1989–1990. Thus, the model might have been a useful and accurate tool for analysts thinking about the potential for ethnically based conflict in Yugoslavia.

There is a potential built-in bias in using the model to evaluate a retrospective case, in that in hindsight it is easy to identify the crucial events in the evolution of the ethnic conflict in Yugoslavia. However, the model specifies strict and objective guidelines about the data to be evaluated. There is room for an individual analyst to make judgments and assessments, but only within the constraints delineated by the model. As such, the above analysis is far from an exercise in retelling a now-familiar story. The framework points to what the analyst should have been looking for, and the parsimony of the approach is one of its assets. So even though the story of the Yugoslav breakup is now well-known, and most specialists subscribe to the notion that Milosevic's harnessing of Serbian nationalism brought about the end of Yugoslavia,[34] the framework applied here allows us to see how the course of events might have been anticipated better.

Perhaps the best use of the model and the analysis contained herein is the clear linkage between specific goals, the resources amassed, and the strategies and choices open to both the mobilized group and the state. There was nothing irrational about the strategies pursued by the federal Presidency or by Milosevic as Yugoslavia slid into collapse. The state's breakdown was tied to the logic of ethnic mobilization, a leader who chose to exploit ethnic attachments by hijacking the administrative machinery of a strong constituent member of the federal state, and the economic, organizational, and coercive resources available to each side. The resource base determined the range of choices between accommodation and strife. The fact that Milosevic used an aggressive set of strategies (including the threat of force) stemmed from the resources he amassed vis-à-vis the federal state. Similarly, the ineffective and weak federal attempts to deal with Milosevic illustrated the fundamental lack of resources available to the state vis-à-vis a militant mobilized ethnic Serbian group.

---

[34]Sabrina P. Ramet, *Balkan Babel: The Disintegration of Yugoslavia from the Death of Tito to Ethnic War,* 2nd edition, Boulder, CO: Westview Press, 1996; see also Silber and Little, op. cit.

# ANNEX:  DEMOGRAPHIC CHARACTERISTICS OF YUGOSLAVIA IN THE LATE 1980s

The following information about population characteristics is based upon the situation in Yugoslavia at the beginning of 1990.  The information presented here is the basic reference for the analysis in this chapter.

**Name:**  The Socialist Federal Republic of Yugoslavia (SFRY).

**Nature of government:**  Modified communist system in a federal state with extensive powers held by the constituent republics; the federal parliament (the Assembly of the Socialist Federal Republic of Yugoslavia) consisted of two chambers:  the Federal Chamber and the Chamber of Republics and Provinces.  The Federal Chamber had most of the lawmaking functions; the Chamber of Republics and Provinces provided an additional check by the republics on the federal government.

**Organization of the state:**  The state comprised six "socialist republics":  Bosnia-Herzegovina, Croatia, Macedonia, Montenegro, Serbia, and Slovenia.  In addition, Serbia contained two "autonomous provinces" (with rights similar to those of republics):  Kosovo and Vojvodina.

**Date of constitution:**  1974.

**Population:**  22,418,331 (1981 census).

**Major ethnic groups:**  Official "nations":  Serbs, Croats, Muslims, Slovenes, Macedonians, Montenegrins; official "nationalities":  Albanians, Hungarians (and nine others, ranging from 10,000–150,000:  Roma, Turks, Slovaks, Romanians, Bulgarians, Ruthenians, Czechs, Italians, and Ukrainians); as well as numerous smaller groups.

**Languages:**   Official languages:   Serbo-Croatian, Macedonian, Slovene (all belonging to the South Slavic language group).   Serbo-Croatian was the dominant language.

**Religions:** Orthodox Christianity, Catholicism, Islam.

**Population statistics.**  The principal ethnic groups were South Slavic: Serb, Croat, Montenegrin, Slovene, Macedonian, and (Slavic) Muslim.  Of the non-Slavic ethnic groups, the Albanians (who formed a majority in the Kosovo province) and the Hungarians (concentrated in the Vojvodina province) were the largest.  Yugoslavia was divided into administrative units based on ethnicity, with each republic named after its eponymous ethnic group.  The exception was Bosnia-Herzegovina (a nonethnic name), which had a population comprised of Serbs, Muslims, and Croats.  In Yugoslavia the term "Muslim," especially when used in official discourse, had ethnic rather than religious connotations.  The majority of Yugoslavia's Muslims resided in Bosnia-Herzegovina and the adjoining part of central-western Serbia (Sanjak).  The two autonomous provinces (Kosovo and Vojvodina) had non-ethnically-derived names (based on historical provinces).

**Table 3.16**

**Population of Yugoslavia by Ethnicity**

| Ethnic Group | Population (in thousands) |
|---|---|
| Serbs | 8,141 |
| Croats | 4,428 |
| Muslims | 2,000 |
| Albanians | 1,731 |
| Slovenes | 1,754 |
| Macedonians | 1,342 |
| Montenegrins | 579 |
| "Yugoslavs" | 1,209 |
| Hungarians | 427 |
| Others | 818 |
| Total | 22,428 |

SOURCE: 1981 census.

Other than possible undercounting of Albanians, and the tendency for any republic or province to have incentives in place for people to claim its principal ethnicity, there appear not to have been any major problems with the accuracy of the census. The self-identification of "Yugoslav" was most commonly selected by offspring of interethnic marriages (1.2 million people declared this to be their nationality in the 1981 census—only just over 200,000 did so a decade earlier—out of a total population of 22.4 million).[35]

Population growth rates varied greatly, depending on the ethnic group. The high fertility rate among the Albanians was responsible for a 23 percent rate of population increase in Kosovo (the poorest region) between 1981 and 1990. Conversely, the wealthiest and developed republics (Slovenia and Croatia) showed the country's lowest rate of population increase.

Table 3.17

Ethnic Groups by Republic and Province

| Republic or Province | Percent of Population by Ethnicity | | | | | | | | |
|---|---|---|---|---|---|---|---|---|---|
|  | Monte-negrin | Croat | Mace-donian | Muslim | Slovene | Serb | Alban-ian | Yugo-slav | Other |
| Yugoslavia | 2.58 | 19.75 | 5.97 | 8.92 | 7.82 | 36.30 | 7.72 | 5.44 | 5.51 |
| Bosnia-Herz. | 0.34 | 18.38 | 0.05 | 39.52 | 0.07 | 32.02 | 0.11 | 7.91 | 1.60 |
| Montenegro | 68.54 | 1.18 | 0.15 | 13.36 | 0.10 | 3.32 | 6.46 | 5.35 | 1.54 |
| Croatia | 0.21 | 75.08 | 0.12 | 0.52 | 0.55 | 11.55 | 0.13 | 8.24 | 3.61 |
| Macedonia | 0.21 | 0.17 | 67.01 | 2.07 | 0.03 | 2.33 | 19.76 | 0.75 | 7.67 |
| Slovenia | 0.17 | 2.94 | 0.17 | 0.71 | 90.52 | 2.23 | 0.10 | 1.39 | 1.77 |
| Serbia Proper | 1.35 | 0.55 | 0.51 | 2.66 | 0.14 | 85.44 | 1.27 | 4.78 | 3.29 |
| Kosovo | 1.71 | 0.55 | 0.07 | 3.70 | 0.02 | 13.22 | 77.42 | 0.17 | 3.14 |
| Vojvodina | 2.13 | 5.37 | 0.93 | 0.24 | 0.17 | 54.42 | 0.19 | 8.22 | 28.33 |

SOURCE: 1981 census.

---

[35]A census was held in the former Yugoslavia in early 1991. Although the data are available, they are not used here, since the tense interethnic situation at the time seems to have biased the results. In ethnically divided societies, a census takes on a highly political role. To paraphrase Horowitz, a census in an ethnically divided society is an election and an election is a census (Donald L. Horowitz, *Ethnic Groups in Conflict*, Los Angeles: University of California Press, 1985). In any event, the aim of this case study is to look at the situation as it existed in January 1990. At that time, only the 1981 census figures were available.

Ethnicity and religion were closely related in Yugoslavia. Slovenes, Croats, and Hungarians have been associated traditionally with Roman Catholicism. The Serbian Orthodox Church has been associated traditionally with Serbs and Montenegrins. Macedonians also have been associated with Orthodoxy (before the establishment of the independent Macedonian Orthodox Church in the 1960s, the religious affiliation of the Macedonians was with the Serbian Orthodox church). Islam was associated primarily with Albanians and Slavic Muslims inhabiting the central part of the country (Bosnia-Herzegovina, southwestern Serbia, and northwestern Macedonia).

Language and ethnicity were also closely related. The language spoken by Serbs, Croats, Montenegrins, and Muslims was essentially the same South Slavic language: Serbo-Croatian (with only dialectic differences between them, primarily pronunciation stress and some vocabulary). The Croats used the Latin alphabet, whereas the others used the Cyrillic alphabet (the Muslims used both, depending on

Table 3.18

Population Distribution of Yugoslavia, 1981–1990 (est.)

| Republics | Land Area $(km^2 \times 1000)$ | Population 1981 (1,000s) | Density (persons per $km^2$) | Population 1990 (1,000s) (estimated) | Density (persons per $km^2$) | Population Change, 1981–1990 (percent) |
|---|---|---|---|---|---|---|
| Serbia | 88.4 | 9,279 | 104 | 9,815 | 111 | 5.46 |
| Serbia Proper | 56.0 | 5,666 | 101 | 5,717 | 102 | 0.89 |
| Vojvodina | 21.5 | 2,028 | 94 | 2,042 | 95 | 0.69 |
| Kosovo | 10.9 | 1,585 | 147 | 2,056 | 189 | 22.91 |
| Croatia | 56.5 | 4,578 | 82 | 4,726 | 84 | 3.13 |
| Slovenia | 20.3 | 1,884 | 90 | 1,924 | 95 | 2.08 |
| Bosnia-Herz. | 51.5 | 4,116 | 83 | 4,795 | 93 | 14.16 |
| Macedonia | 25.7 | 1,914 | 73 | 2,193 | 85 | 12.70 |
| Montenegro | 13.8 | 583 | 43 | 664 | 48 | 12.20 |
| Total | 255.8 | 22,354 | 87 | 24,117 | 94 | 7.31 |

SOURCES: Statistical Yearbook of the Socialist Federal Republic of Yugoslavia; Taken from the Economist Intelligence Unit (EIU) Yugoslavia Country Profile, August 27, 1991.

their geographical residence).  Slovene and Macedonian were distinct South Slavic languages related to Serbo-Croatian.  Slovenes used the Latin alphabet, Macedonians used the Cyrillic alphabet.  Albanians and Hungarians used their own, non-Slavic languages.

Serbia was the most populous of the Yugoslav republics.  Including the autonomous provinces of Vojvodina and Kosovo, Serbia's population was roughly 10 million—almost twice as large as the second most populous republic, Bosnia-Herzegovina—and 40 percent of Yugoslavia's total population.  In addition, Montenegrins were seen as close "ethnic cousins" of the Serbs.

Slovenia was the most ethnically homogenous republic, and Bosnia-Herzegovina was the most diverse (with no one ethnic group having a majority).  Besides forming the majority of population of Serbia proper and Vojvodina, Serbs made up significant minorities in Bosnia-Herzegovina, Kosovo, and Croatia.  In addition, the Serbs living outside of Serbia tended to be concentrated geographically, placing them in a position of local ethnic majority group.

# Chapter Four

# THE SOUTH AFRICAN RETROSPECTIVE CASE

## Pearl-Alice Marsh and Thomas S. Szayna

## INTRODUCTION

This chapter applies the "process" model for anticipating the incidence of ethnic conflict to the case of the end of *apartheid* and the peaceful transition of power in South Africa in 1990–1994. It examines the case of South Africa from the perspective of what an intelligence analyst might have concluded about South Africa's propensity toward ethnic violence had she used the "process" model to examine the country's situation in early 1989.

In essence, this chapter examines the mobilization of the majority "black" population in South Africa by the African National Congress (ANC) and the attempt by this mobilized group to move South Africa toward an inclusive political system. As such, the conflict in question is between the ANC and the *apartheid* South African state. The primary time frame of interest is the period 1985–1989. As in the Yugoslav case, the analytic orientation is in line with the "process" model's focus on one particular kind of ethnic conflict, namely, the mobilization of an ethnic group challenging the state.

The choice of the date stems from two reasons. One, the cutoff date comes before the major shift away from *apartheid* but after a sustained period of tinkering with the system (1983–1989) that introduced the possibility of major change. An earlier cutoff date, for example mid-1988 or even mid-1987, would not have led to different results but would have prevented the inclusion of several important factors that drove the process of accommodation. Two, the time frame is realistic in that the fate of South Africa (ranging anywhere from peaceful transition, to muddling-through, to collapse into violent civil war) was not preordained in early 1989. Many factors, internal and external, might have changed the outcome. Once the *apartheid* state legalized the ANC in 1990, and especially after the setting up in 1991 of a mechanism to negotiate the transfer of power, the likelihood of a negotiated transfer of power increased, but even in late 1993 there was nothing inevitable about the outcome.

As was the case with Yugoslavia, some may argue that a simple group-versus-state model is not realistic when applied to the multiethnic conditions of 1980s South Africa. Perhaps a model that took greater account of the substantial intergroup competition in South Africa (intra-"black" divisions, inclusion of the "coloureds" or "Asians," looking at the "whites" as subdivided into Anglophones and Afrikaners) would be a more accurate portrayal of the situation. While the various groups inhabiting South Africa certainly were not monolithic (and every group has subgroup characteristics), the state's policy of using "racial" criteria to define individuals formed the basis for social divisions in South Africa. In other words, individuals whose self-identity was Xhosa, Zulu, or Tswana, even if they felt distinctly different from each other and spoke different languages, were still lumped together as "blacks" by the state, and none had political rights under *apartheid* South Africa. The state's creation of "homelands" for the black majority population recognized ethnolinguistic differences among the "blacks," but the "homelands" provided no political rights for them in the context of the highest governmental bodies of the state and politically only served to divide the black majority population. Similarly, even though Anglophone and Afrikaans-speaking individuals with European physical features may have felt little in common with each other, the state lumped them together as the politically privileged "whites." The main polyethnic categories of "whites," "blacks," "coloureds," and "Asians" were creations of state policy in South Africa, even though they may have built on earlier informal social distinctions. But the creations became real and political bargaining came to be carried out under these categories. Thus, the authors do not see that the added benefit of any complicated multigroup model would offset the exponentially greater complexity of the model. Parsimony is a crucial element of any model, and narrowing a conflict to its most essential aspects is the goal of the "process" model.

This chapter uses the terminology of ethnic discourse in South Africa because, as explained above, the polyethnic categories created by the state became a political and social reality.  The very terminology of "whites," "blacks," "coloureds," and "Asians" may be offensive, as it is implicitly based on a racist mindset.[1]  The use of *apartheid* terminology is in no way meant to signify the acceptance of such terms or of the mindset that led to their construction.

Some may further question the applicability of a model designed to analyze "ethnic" conflict to the racially polarized situation in South Africa.  However, in line with Horowitz,[2] the authors see "race" as a social construct and a rigid subcategory of ethnicity.  As many have noted, race is not a preexisting category; it "is not found, but 'made' and used."[3]

Following the format of the previous chapter, four sections follow this introduction.  The first section examines the structure of closure and provides an analysis of which ethnic groups were privileged and which were dominated (the demographic characteristics of South Africa in the late 1980s, on which the analysis is based, is appended at the end of the chapter).  Then it examines the strength of the challenging ethnic group—the black majority represented primarily by ANC—by looking at its mobilization process.  The second section looks at the capabilities the state—the Republic of South Africa— could have brought to bear in dealing with the challenging group. The third section examines the strategic choices, arrived at on the basis of the assessments in earlier sections, that the state and the group were likely to pursue vis-à-vis each other, given their resource base group.  Finally, a few observations on model application conclude the chapter.

---

[1]Alternative terms are also deficient for the purposes of the analysis contained herein. For example, the term "Africans" to denote "blacks" (or more properly, indigenous Africans) has a confusing meaning, since it denies the "African-ness" of individuals with physical features associated with indigenous Europeans but who can trace ancestry for ten generations in southern Africa.  Similarly, the term "mixed-race" to denote "coloureds" is no improvement.

[2]Donald L. Horowitz, *Ethnic Groups in Conflict*, Los Angeles:  University of California Press, 1985.

[3]Anthony W. Marx, "Race-Making and the Nation-State," *World Politics*, Vol. 48, January 1996, p. 180.

## ASSESSING THE POTENTIAL FOR STRIFE

### Closure in the Political and Security Realms

In terms of closure in the political realm, the president and a tricameral parliament held legislative power under the terms of the 1984 constitution. The three houses of the parliament were segregated on the principle of racial categories. The white population was represented by the House of Assembly, with 178 members (166 elected directly and 12 indirectly). The coloured population was represented by the House of Representatives, with 85 members (80 elected directly and 5 indirectly). The Asian population was represented by the House of Delegates, with 45 members (40 elected directly and 5 indirectly). Each house legislated its own community's affairs (in other words, the House of Assembly dealt with all issues within the white community). "Own affairs" were defined as matters which specially or differentially affected a population group in relation to the maintenance of its identity and the upholding and furtherance of its way of life, culture, traditions, and customs.

The three houses nominally shared responsibility for national affairs, though the House of Assembly held most of the power. It was up to the president to decide whether a specific issue was a national or a community affair. All three houses and the president had to approve legislation concerning national affairs. If a consensus among all three houses was not achieved, laws still could be enacted through approval by the President's Council and any one of the houses. The Supreme Court did not have the power of judicial review of parliamentary legislation.

The parliament elected the president to a seven-year term by means of an electoral college. Through a majority-voting process, the three houses chose 50 white, 25 coloured, and 13 Asian members, respectively, as members of the electoral college. It was up to the president to initiate legislation. The president was assisted by the President's Council, composed of 60 members (20, 10, and 5 members nominated by the three houses, respectively; 15 members nominated by the president; and 10 members nominated by the opposition parties). The president also appointed a Ministers' Council for each house, from the house's majority party. The Ministers' Councils carried out an administrative role for the specific "racial" group. The president also chose his own Cabinet, primarily from members of the three houses.  The president had the right to suspend the parliament and to dissolve the House of Assembly.

The president was also vested with control and administration of black affairs.  Governance of black affairs, in general, was carried out by homeland governments.  Homelands were set up on the basis of the 1959 Promotion of Native Lands Self-Government Act, which envisaged eventual political autonomy for blacks within South Africa.  Each of the ten homelands had a legislative assembly and an executive council, headed by a Chief Councillor.

The constitution established a strong presidency and made provision for a nominal representative role in national affairs for the coloured and Asian populations.  But the constitution made certain that the white representative body could overrule both the coloured and Asian parliamentary chambers and, if necessary, elect the president and pass laws based simply on the majority party of the white chamber of parliament.  In addition, the constitution made no provision for national political participation by blacks residing in South Africa.

These patterns are evident from a closer look at the highest-ranking individuals of the political apparatus in South Africa.  The Presidency, the Cabinet, and the President's Council were the most influential institutions concerning national affairs.  The dominant political party, the National Party (NP) retained control of the House of Assembly during the elections in May 1987.  The NP's chief source of support came from Afrikaners (though increasingly by the late 1980s, Anglophones too voted for the party).  Table 4.1 shows the ethnic composition of the politically highest-ranking individuals.  Although the nonwhite parliamentary chambers played a minor role in national affairs, the chairmen of these bodies could (and did) play a prominent role in channeling the concerns of the group to the president (see Table 4.2).

Provincial and homeland administrations had direct say over local matters, but because of the segregation of representative bodies along racial lines and the concentration of power for national policy with the Presidency, the latter held a virtual monopoly over executive and legislative powers regarding the main paths of national development.

Table 4.1

South African Presidency and Main Cabinet Members, April 1989

| Position | Name | Ethnicity |
|---|---|---|
| President | P.W. Botha | white, Afrikaner |
| Minister of Finance | B.J. du Plessis | white, Afrikaner |
| Minister of Administration and Privatization | D.J. de Villiers | white, Afrikaner |
| Minister of Economic Affairs and Technology | D.W. Steyn | white, Afrikaner |
| Minister of Home Affairs and Communications | S. Botha | white, Afrikaner |
| Minister of Constitutional Development and Planning | J.C. Heunis | white, Afrikaner |
| Minister of Justice | H.J. Coetsee | white, Afrikaner |

SOURCE: *Defense and Foreign Affairs Handbook*, Alexandria, VA: International Strategic Studies Association, 1989.

Table 4.2

South African Chairmen of Ministers' Councils, April 1989

| Chairman of Ministers' Council | Name | Ethnicity |
|---|---|---|
| Council for White Own Affairs | C.V. van der Merwe | white, Afrikaner |
| Council for Coloured Own Affairs | A. Hendrickse | coloured |
| Council for Indian Own Affairs | A. Rajbansi | Indian |

SOURCE: *The Statesman's Yearbook*, 1988–89.

Since cabinet ministers had the authority to appoint senior political appointees as heads of departments and directorates, the composition of the director-level management of the South African bureaucracy mirrored that of its highest-level officials (see Table 4.3). Accordingly, because the NP controlled the parliament and the Cabinet, NP loyalists predominated in the executive machinery of the government.

In terms of closure in the security realm, the constitution centralized authority over security matters with the office of the Presidency and, in comparison with the pre-1984 constitution, withdrew such matters from parliamentary oversight. The president also had the right to define which matters fell in the security realm. The president chaired the State Security Council, the top state body for security issues, and was the nominal chief of the armed forces.

**Table 4.3**

**South African Director-Level Managers, 1986**

| Position | Name | Ethnicity |
|---|---|---|
| Dircctor-General, Finance | J.H. de Loor | white, Afrikaner |
| Director-General, Trade and Industry | J.P. du Plessis | white, Afrikaner |
| Director-General, Minerals and Energy Affairs | L. Albert | white, Afrikaner |
| Director-General, Home Affairs | G.B.S. van Zyl | white, Afrikaner |
| Director-General, Constitutional Development | A.H. van Wyk | white, Afrikaner |
| Director-General, Public Works and Land | P.C. van Bloommestein | white, Afrikaner |

NOTE: Data for April 1989 are not available. However, the 1986 data approximate closely the composition of director-level managers in 1989, as the same party and virtually the same government held power.

SOURCE: *Official Yearbook of the Republic of South Africa,* Pretoria: South African Communication Service, 1986.

The control of the Presidency by the dominant party in the House of Assembly meant a virtual monopoly for the NP on security issues and the highest-ranking security personnel. The minister of defense controlled the administrative machinery of the armed forces. The chief of the armed forces—the South African Defense Force (SADF)—had operational control over the military. The Ministry of Law and Order controlled the South African Police. As with the top political appointees in the administrative machinery of the state, the pool of candidates for top civilian positions in the security realm came from NP loyalists. These were almost exclusively Afrikaners (see Table 4.4). In the police and the armed forces, the association of *apartheid* with the NP appears to have led to self-selection and promotion patterns that overwhelmingly favored Afrikaners. A survey of top military, police, and intelligence apparati throughout the 1980s shows that these areas were almost an exclusive domain of the white Afrikaners. In the armed forces, Afrikaners comprised 85 percent of the officer corps.[4]

---

[4]Colin Bundy, "History, Revolution and South Africa," *Transformation,* Vol. 4, 1987, pp. 60–75.

Table 4.4

Top South African Security Authorities, 1988–1989

| Position | Name | Ethnicity |
|---|---|---|
| President | P.W. Botha | white, Afrikaner |
| Minister of Foreign Affairs | R.F. Botha | white, Afrikaner |
| Minister of Defense | General M. Malan | white, Afrikaner |
| Chief of Staff, SADF | Lt. General J.J. Geldenhuys | white, Afrikaner |
| Minister of Law and Order (police) | A. Vlok | white, Afrikaner |
| Chief, South African Police (SAP) | General Hennie de Witt | white, Afrikaner |
| Chief of National Intelligence Service (NIS) | Niel Barnard | white, Afrikaner |

SOURCE: *Defense and Foreign Affairs Handbook*, Alexandria, VA: International Strategic Studies Association, 1989.

**Assessment of closure in the political and security realms.** Under the guise of a democratic set of rules (even if the democratic system was only for a minority of the people), the political system in South Africa in fact established few limitations on the power of the single most powerful political organization of the whites. Because of the strong in-group proclivities and numerical prominence of the Afrikaners among the whites, the Afrikaners—through the vehicle of the NP—had a hold on political power in South Africa.

Similar patterns of dominance are observable in both the political and security realms. There is no question that the Afrikaners had a complete hold on all positions of political power, at both the highest executive and top administrator levels. Even the other white groups appear to have been shut out from these positions. The lack of presence of the coloureds and the Asians was total. Most of all, the majority blacks did not have access even to the institutions of national political power.

The same pattern was observable in the security apparatus. Although rank-and-file membership of the military and security apparati differed little from the general makeup of the white population (service was compulsory for white citizens only), discrepancies (in favor of the Afrikaners over the Anglophone whites) appeared at the middle levels. At the highest levels, the dominance of the Afrikaners was nearly complete.

In short, the static South African system of *apartheid* had no provision for a dynamic for change. Restrictions on access to the political power system were based on "racial" distinctions, codified by law, and enforced through the machinery of the state. The *apartheid* system severely limited access to political power for all groups other than whites (and especially for blacks) by providing legal roadblocks to participation in politics. The black population had virtually no rights in the political system, other than a limited "self-governing" role in the homelands. The coloured and Asian populations had a nominal input into the national political process, but their role was constrained by institutional rules. Even among the politically privileged whites, rules on parliamentary procedure made it difficult for the Anglophone population to challenge the Afrikaners, if the latter acted in a unified manner.

In theory, there existed a process for the replacement of elites within the political hierarchy. But in fact there was little potential for peaceful change by nonwhites, since access to the elite level was limited to whites. The collectivist principle of race was by far the most important determinant of not only access but even participation in the political process. A whole legal edifice, backed up by a substantial use of force and suppression of any organized dissent, supported the constraints.

Access to positions of power and the ability to change the power structure varied greatly according to the racial categories. In theory, the strong Afrikaner position was most threatened by other whites, for a fundamental political realignment in the House of Assembly and the replacement of the NP by another party spanning the divisions among whites could have reduced substantially the Afrikaner domination of all top positions of political power. But in practice such a realignment was highly unlikely, and as long as the Afrikaners were politically unified, they had a stranglehold on the House of Assembly and political power in the country. The NP government tolerated criticism by other whites and generally lived up to democratic norms within the white community. For example, the leadership of the Progressive Federal Party (PFP) used its position in the House of Assembly to challenge the government on a wide range of policies, though because of PFP's distinct minority status, its impact amounted to little more than open voicing of opposition.

Relative to the whites, the coloureds and Asians had a distinctly inferior position in the political system. Although the 1984 constitution
gave the two groups some access to political power, their practical
role was limited to the ability to voice opinions in the governing process. Because of the institutional rules of the parliament and the
segregation at the level of parliamentary representation, even if the
House of Representatives and House of Delegates were unified in
rejection of the dominant party in the House of Assembly, they
would still have no significant input into policymaking. In addition,
the NP did not extend the same level of tolerance to criticism from
the coloured and Asian opposition in the parliament that it extended
to the whites (for example, H.J. Henrickse, Chairman of the Ministers' Council of the House of Representatives, was forced to apolo-
gize to the president for offending him during one of Henrickse's
challenges).

But the blacks were, by far, in the worst position. Having no representation at the parliamentary level, and with a policy in place
designed to take away gradually even their South African citizenship
(the homelands policy),[5] they had absolutely no direct influence over
or access to the political system. Outside the homelands, the only
governing role for blacks was that of participation in town councils,
located in townships and responsible for collecting rents and service
fees. The government did not tolerate any organized black political
opposition, and virtually all such organizations were banned. These
extralegal organizations included the African National Congress
(ANC), Pan Africanist Congress (PAC), Azanian Peoples' Organization
(AZAPO), Black Peoples' Convention (BPC), and black trade unions.

## Assessing Closure in the Economic Realm

The distribution of wealth in South Africa was highly skewed along
racial lines, with the whites having an inordinate share. The hierarchy evident in the political sphere was replicated in the economic

---

[5]In 1985, President Botha and the NP government conceded that the policy of
automatic disenfranchisement of blacks was not prudent. The government declared
that no one would be disenfranchised simply because their ethnicity qualified them
for citizenship in one of the homelands. However, the government did not extend the
rights of full citizenship to blacks residing outside the homelands.

sphere, with the whites in an unchallenged position at the top, coloureds and Asians in the middle, and blacks at the bottom. Among the whites, both the Afrikaner and Anglophone populations were privileged.

Without a doubt, the wealthiest strata of the population were composed entirely of whites, both Afrikaner and Anglophone, though Anglophones seemed to dominate.  The top corporate officials of major South African corporations (banking, manufacturing, commerce, and mining—especially diamonds and gold) were probably among the wealthiest individuals.  The monopoly position of the major conglomerates (Barlow-Rand, Anglo-American, De Beers, and South African Breweries) enhanced their profitability.

At the general population level, income was inequitably distributed, with the primary differences along racial lines.  The general pattern of occupation by racial categories amounted to a hierarchy, with blacks primarily in unskilled jobs, coloureds and Asians in semiskilled jobs, and whites in skilled positions.  The roles of owner, manager, and technician were an almost exclusive domain of the whites.  Although there were some unemployed whites, there was no white poverty, while the majority of nonwhites lived in poverty and destitution.  Virtually all white households employed domestic workers, and many employed a full range of servants, maids, gardeners, and cooks (so that sometimes there were more domestic workers than related members of the white household).  Automobiles, telephones, televisions, and computers were luxuries owned primarily by whites.  By the 1980s, the previous income disparities between Afrikaners and Anglophones (with Afrikaners in a less privileged position) had largely dissipated.

As Table 4.5 shows, with subsistence level for a family of six calculated at R354/month,[6] most blacks lived in poverty and barely above subsistence level.  A small urban black middle class did exist, made up of teachers, doctors, lawyers, nurses, and undertakers, as well as employees of white-owned firms, particularly public relations and human resources, but also in small business and industrial consult-

---

[6]*Survey of Race Relations—1986,* Johannesburg:  South African Institute of Race Relations, 1986.

**Table 4.5**

**Average Monthly Earnings in South Africa, by "Racial" Categories
(Rand/month)**

| Year | Black | Coloured | Asian | White |
|------|-------|----------|-------|-------|
| 1983 | 310   | 416      | 586   | 1,211 |
| 1984 | 373   | 494      | 690   | 1,405 |
| 1985 | 418   | 561      | 770   | 1,561 |

1 South African rand (1986) = US$ 0.49 (1986).

SOURCE: *Survey of Race Relations—1986,* Johannesburg: South
African Institute of Race Relations, 1986.

ing firms. However, such people constituted an extreme minority
among the black population.

**Assessment of closure in the economic realm.** There is clear evidence that most whites were in the top 20 percent of the population
in terms of wealth, whereas virtually all blacks were in the lower 60
percent of the income distribution scale. Moreover, the distribution
curve appears to have been S-shaped, with a steep drop-off after the
first 20–25 percent.

Was there a mechanism available for changing the static snapshot
presented above? Clearly, a developed capitalist system was in place
in South Africa and capital accumulation and individual initiative
were encouraged, with the major proviso that different races had
very different levels of preparation for the process. Moreover, informal practices as well as some legal restrictions gave advantages to
the whites and penalized blacks the most.

By 1986, specific laws no longer precluded blacks and other nonwhites from economic aspirations. Though still severely underprivi-
leged, urban blacks could become economically successful. But for
the vast majority of nonwhites, and blacks especially, poverty, poor
education, lack of capital, and hostile governmental agencies made
accomplishing the task difficult. The problem stemmed from the
legacy of several decades of longstanding legal discrimination that
included even the reservation of jobs for whites. Removing the legal
discrimination did not address the legacy of the far-reaching

discrimination or the likelihood that the same economic inequalities would be kept intact through informal mechanisms.

Until the mid-1980s, *apartheid* laws had made it virtually impossible for blacks to accumulate capital. The 1963 regulations implementing the 1957 Native Law Amendment Act created the following conditions: (1) blacks were forbidden to run more than one business at a time; (2) blacks had to have the right of "permanent" residence in a city before they could function as a business entity (something virtually impossible bureaucratically); (3) blacks were forbidden to form joint-stock companies or companies of more than one person; (4) blacks were forbidden to start up banking or financial businesses, manufacturing industries, or wholesale trading businesses; (5) blacks were not allowed to own business premises. In addition, the Group Areas policy restricted the access of blacks to housing and commercial land in black areas. The Group Areas Act controlled the acquisition of immovable property between individuals or companies of the various racial groups and controlled the occupation of land, buildings, and premises for all purposes. Assuming that wealth is generated primarily by the ownership and investment of capital, the earlier laws had virtually precluded black access to capital, property ownership, and wealth.

Remaining legal prohibitions as well as informal constraints certainly played a role in thwarting the economic mobility of blacks, but the structure of access to education and, consequently access to economic opportunities, was most telling of the lack of preparation for economic success. As Table 4.6 indicates, in terms of educational expenditures, there was a clear hierarchy along racial lines, with whites at the top and blacks at the bottom.

Moreover, government policies requiring school fees prohibited access to effective training in business development and management. Only in the informal sector (as hawkers, small shop proprietors, and as laborers) were blacks economically engaged. There were restrictions on black economic activity even in the "self-governing" homelands, for only through the Native Lands Investment Corporations, heavily subsidized by the South African government, were blacks allowed to build businesses in the homelands.

Table 4.6

Per-Capita South African Governmental Educational
Expenditures by "Racial" Group, 1985–1986

| Group | Expenditures (in rand) |
|---|---|
| Blacks, South Africa proper | 387.02 |
| Blacks, homelands average[a] | 243.92 |
| Coloured | 891.62 |
| Asian | 1,386.00 |
| White | 2,746.00 |

[a]Based on data for six homelands. These ranged from a
high of R277.70 in Bophuthatswana to a low of R184.00 in
Lebowa.
SOURCE: *Survey of Race Relations—1986,* Johannesburg:
South African Institute of Race Relations, 1986.

## Closure in the Social Realm

Status distinctions in South Africa were abundantly clear, crucial in
all aspects of public interaction, and legally upheld through the state
machinery. Status stemmed primarily from outward physical features, generalized and categorized into "racial" categories based on
an ideological racism. A status stratification map for South Africa
(taking into account only the "official races") might run along the
following lines (see Table 4.7).

The basis for the pattern of status and stratification in South Africa
stemmed from race, vocation, and national origin. However, race
was by far the most important determinant, and a racial hierarchy
was established in the constitution.

The doctrine of *apartheid* ("apartness"), or separate development,
was the ideological underpinning for a whole system of segregationist laws, a hierarchy of groups, and a system of privileges and discrimination. The *apartheid* doctrine stemmed from traditional segregationist South African (primarily Afrikaner) practices legitimized
by doctrines of the Dutch Reformed Church and had, at its core, an
all-consuming belief in white superiority and the need to safeguard

Table 4.7

Status Stratification

| Group | Status |
| --- | --- |
| Whites | + + |
| Coloureds | – |
| Asians | – |
| Blacks | – – |

+ + = high status; – – = low status
(compiled on the basis of data
presented earlier).

white culture from the indigenous Africans.[7]  An enormous consciousness of outward physical differences and the "scientific" race theory of the late 19th and early 20th centuries formed the basis for the belief that various ethnic groups are fundamentally distinct, have different destinies, and must evolve in separation from each other. Christian nationalism in South Africa reconciled the conflict between the Christian monogenesis of humankind and race theory by constructing explanations based on cultural essentialism.[8]

The *apartheid* doctrine built on a highly developed identity of a "chosen people" among the Afrikaners.[9]  Based on myths of the Afrikaners' trek into South Africa's heartland and their wars against the British and native Africans, a set of beliefs emerged that the Afrikaner "volk" were destined by God to conquer both the land and the natives to prosper, protect, and preserve the white race.  Underpinning these beliefs was a strong sense of distinctness from and an

[7]For an excellent comparative look at the construction and meaning of race in South Africa among whites, see Gerhard Schutte, *What Racists Believe: Race Relations in South Africa and the United States,* Thousand Oaks, CA: Sage, 1995.

[8]Saul Dubow, *Scientific Racism in Modern South Africa,* Cambridge:  University of Cambridge Press, 1995.

[9]Gerhard Schutte, "Afrikaner Historiography and the Decline of Apartheid:  Ethnic Self-Reconstruction in Times of Crisis," in Elizabeth Tonkin, Maryon McDonald, and Malcolm Chapman (eds.), *History and Ethnicity,* London and New York:  Routledge, 1989, pp. 216–231.

imposition of an imagined identity upon the indigenous African, describing him as the essential "other" to the European.[10]

In the Afrikaner ideological framework, there was no room to imagine native Africans as intellectual equals or capable of understanding Western ideas, for Afrikaners saw the different levels of development between the Europeans and native Africans as genetically based. Within such a mindset, it followed that native Africans were not culturally, socially, or intellectually fit to contribute to South Africa's development and needed only a basic education so as to be allowed to work in unskilled jobs. A paternalistic attitude toward the native Africans as well a belief that they were docile and happy and had no political ambition permeated the Afrikaner mindset. For Afrikaners, the indigenous African population was not monolithic but consisted of many different ethnic and "tribal" groups. But it was up to the highest-developed tribe, the "white tribe" of Afrikaners, to rule over the amalgam of black, coloured, Asian, and white tribes.

In theory, *apartheid* meant the separate but equal development of all "racial" groups. But the emphasis on "separate" led to the formulation of a whole range of laws designed to ensure segregation. Although the accession of the NP to power in 1948 marked the full application of *apartheid* in the political realm, earlier moves had set the stage for the separation of races. In 1913, just three years after the founding of the Union of South Africa, the Native Land Act introduced the principle of territorial separation of black and white populations by fixing the boundaries of tribal lands and prohibiting acquisition of land by blacks outside these boundaries. The Native Trust and Land Act (1938) pursued the policy of geographically distinct territories for blacks by designating the native African homelands. The Immorality Act (1927) banned sexual relations between members of different "races," and the Black Representation Act (1936) removed blacks from the voters' roll. But after the NP political victory in 1948, a whole host of basic segregationist laws followed in 1949–1953. The Group Areas Act (1949) gave legal force to the existing practice of segregation in housing, the Population Registration Act (1950) began the process of classifying all inhabitants according to the four "racial" categories (white, black, coloured, Asian), the Pass

---

[10]Saul Dubow, *Scientific Racism in Modern South Africa*.

Laws Act (1952) required blacks to carry internal passports and made entry into urban areas subject to police permission, and the Reservation of Separate Amenities Act (1953) established separate public facilities for blacks and whites.[11]

The homeland policies followed from the implementation of *apartheid* by creating autonomous homelands that eventually were to become independent states. The Promotion of Native Land Self-Government Act (1959) provided for the establishment of such self-governing states on the basis of tribal affiliations, and the Black States Constitution Act (1971) authorized their self-government. The whole thrust of the policy was to remove blacks from even having South African citizenship while keeping them in "reservation"-like areas. The end goal of the policy would have been a division of much of rural South Africa into many "independent" states based on tribal identity, while the urban and commercial farm areas remained a part of the white state. In the 1950s, the many blacks residing in urban areas had no political rights whatsoever; in the 1960s and 1970s, these urban blacks were treated as citizens of their respective homelands rather than the townships in which they resided. The homeland blueprints were suspended and the policies reversed in 1986, with the repeal of the pass laws and the return of South African citizenship to citizens of the four "sovereign" homelands who resided and worked in South Africa proper. Although the government abandoned the homelands idea in 1985–1986 and began to look for a way to integrate urban blacks into the political realm, the overall aim of the homelands policy shows clearly the low status accorded to the blacks. At its height, the policy led to outright suspension of all rights of a large number of blacks in South Africa by taking away their South African citizenship.

Restrictions on coloureds and Asians varied in intensity but never approached the level of legal measures applied to the black population. Both groups were seen as at a higher level of development than

---

[11]The association of *apartheid* with the Afrikaners was clear to all in South Africa. Despite the privileges that Anglophone whites enjoyed as a result of *apartheid,* empirical data shows that hostility among blacks was directed almost entirely against Afrikaner whites, as opposed to all whites. John Duckitt and Thobi Mphuthiy, "Political Power and Race Relations in South Africa: African Attitudes Before and After the Transition," *Political Psychology,* Vol. 19, No. 4, December 1998, pp. 809–832.

the blacks, though not equal to the whites. Rather than in some type of homelands, the coloureds and the Asians were envisioned to live in "white" South Africa, though in segregated townships. With the passage of the constitution in 1984, the two groups even received a nominal power-sharing role.

Among the whites, informal and relatively minor status distinctions ran primarily along Afrikaner/Anglophone lines, with the former seeing themselves as having more of a right to hold power because of their position as the original white settlers of South Africa and their development as a "white tribe" in Africa.

The *apartheid* doctrine amounted to an entire mindset, and the racial hierarchy and status distinctions attendant to it extended to foreigners visiting or residing in South Africa. European foreigners were given identical recognition as white South Africans. Japanese visitors were considered "honorary white," based on lucrative economic relations between South Africa and Japan, but all other Asian foreigners were classified "Asian" and ranked with South African Indians. Affluent foreign Africans and other blacks from outside the southern region had "white" privileges, subject to local custom. African-Americans were granted "honorary white" status to distinguish them as "modern blacks"—different from indigenous Africans—and, in a peculiarly self-contradictory fashion—to affirm the "nonracial" nature of the *apartheid* racial separation policy.

What were the implications of this racially based status distribution? Racial categorization was the single most important consideration for access to a whole range of benefits. Separate rules curtailing freedom of movement, assembly, residence, employment, and organization governed behavior according to one's race.[12] Benefits and restrictions were ever-present and, because of the Reservation of Separate Amenities Act, extended to facilities such as beaches, public transportation, schools, recreational centers, etc. "White only" signs were present at all such facilities. Beginning in 1986, the rules began to be relaxed and some areas began to be opened to all races. But the

---

[12]For a detailed look at the extent of organization of separation between the races, from the administrative boundaries of the state (homelands) to urban planning, to the layout of post offices and households, see A. J. Christopher, *The Atlas of Apartheid*, London and New York: Witwatersrand University Press, 1994.

relaxation of formal rules did not mean the disappearance of informal biases and perceptions that had arisen around the status hierarchy that had been in place for so many years. Many whites probably continued to perceive blacks (and coloureds and Asians) as backward, irrational, and at a lower stage of development. The biases tended to be self-reinforcing and undoubtedly affected myriad everyday decisions, leading to still real, even if informally apportioned restrictions on status and benefits.

Was there a mechanism for change within the South African social stratification system? Status was tied directly and intimately to race, one's race was determined and registered at birth, and there was no real possibility of movement between the racial groups.[13] The political system created an elaborate way of managing intergroup relations, based on segregation and minimum contact. A basic collectivist outlook of treating individuals only as members of substate-level groups underpinned the whole range of *apartheid* laws.

As long as the hierarchy defined in the constitution remained in place, any change in status distribution would be gradual at best. As a society legally stratified by race, mobility for most nonwhites was virtually impossible. An extremely small number of affluent Indians, coloureds, and even blacks existed and had limited access to social amenities reserved primarily for whites. But such people were very few, and the severe constraints on accumulation of wealth—especially for blacks—virtually precluded all but the slowest change in status. In any event, despite the legal changes to the *apartheid* laws, the ideological edifice of the *apartheid* structure remained in place even in the late 1980s.

## Overall Assessment of Closure

Based on the information presented above, Table 4.8 summarizes the degree of closure (in an overall sense as well as in the political, economic, and social realms) experienced by South Africa's main groups in early 1989. To reiterate, closure in Weberian terms refers to the

---

[13]One way of movement between racial groups was by way of formal application and provision of proof that the racial classification was in error. A government body then decided the case.

Table 4.8

Patterns of Closure by "Race" in South Africa

|  | Political | Economic | Social | Overall |
|---|---|---|---|---|
| Whites | Low | Low | Low | Low |
| Coloureds | Moderate | Moderate | Moderate | Moderate |
| Asians | Moderate | Moderate | Moderate | Moderate |
| Blacks | High | High | High | High |

"process of subordination whereby one group monopolizes advantages by closing off opportunities to another group." In the table, a group experiencing a "low" degree of closure has the most opportunities open to it. A group experiencing a "high" degree of closure has opportunities largely closed off.

As noted earlier, an all-encompassing system of ideological racism underpinned closure patterns in the political, economic, and social realms. There was no hypocrisy or ambiguity about the pattern of closure, and the pattern was upheld by the machinery of the state. By virtue of law, South Africans received privilege, status, wealth, and opportunity solely on the basis of race.

Some distinctions existed among the whites (with the Anglophones facing some closure in the political realm) and among the blacks (with urban blacks facing less closure in the political and economic realms than rural blacks), but such distinctions paled in comparison with the sharp distinctions along racial lines. Since there was virtually no potential for change in the stratification patterns (in all three realms), the closure pattern was extremely rigid.

A final ranking of groups along the lines of privileged to dominated—in relative terms—is illustrated in Table 4.9. The specific placement of groups on the privileged-dominated scale is not evenly spaced. In other words, the range of difference between the top two groups was almost insignificant relative to the wide gap separating the blacks from all other groups or the coloureds from the whites as a whole.

Based only on relative deprivation, the blacks of South Africa seem to have had the deepest grievances and the most reasons for seeking to

**Table 4.9**

**Ranking of Groups in South Africa**

| Privileged | White Afrikaners |
| --- | --- |
| | White Anglophones |
| | Coloureds |
| ↓ | Asians |
| Dominated | Blacks |

change the status quo.  The coloureds were the most privileged among the nonwhite groups, though their privileges did not differ all that much from the Asians.  The whites were a group that was the target of all the other groups' grievances.  However, the Asians and the coloureds also had much to lose by treating their position vis-à-vis the whites in zero-sum terms, as the whites could then increase the pattern of closure they faced.

## TRANSFORMING POTENTIAL STRIFE INTO LIKELY STRIFE

This section analyzes retrospectively the process of ANC mobilization in the late 1980s by focusing on the five aspects of mobilization: incipient changes, galvanizing "tipping" events, leadership, resources and organization, and the foreign element.

The reader should keep in mind that militant black (or primarily black) organizations attempting to change the status quo through violent and nonviolent means date back to the founding of South Africa.  The main organization attempting to change the pattern of closure faced by blacks in South Africa, the ANC, was founded in 1912.  The ANC engaged in essentially a civil rights struggle, seeking the extension to blacks of the same rights given to whites under the 1910 constitution.  After the accession of the NP to power in 1948, the ANC transformed into a mass-based political movement that engaged in nonviolent but illegal activities.  In 1961, after a bloody suppression by the police of a peaceful march at Sharpeville, the ANC formed a military arm, the *Umkhonto we Sizwe,* and in 1969 the ANC endorsed the use of violence against the regime (though it shied away from a reliance on armed struggle and looked upon violence as one of the many ways to pressure the government into changing its *apartheid* policies).  Following a bloody police suppression of youth riots (that turned into an uprising) in Soweto in 1976, the ANC launched a guerrilla war in South Africa (primarily sabotage).  In 1984, the ANC extended its guerrilla campaign onto "civilian" targets in addition to the previous emphasis on "hard" targets (police, military, and government installations).  The move proved to be a disaster in terms of international and domestic support, and the policy was retracted.  Armed guerrillas were trained and equipped at sanctuaries in neighboring countries and infiltrated into South Africa.

Thus, considerable violence was already the norm in South Africa by the late 1980s.  Although the Soweto uprising in 1976 was a turning point, a mass mobilization of the black population did not really take place until 1985, with the ANC's successful call to make the black townships ungovernable.  Even then, violence was largely localized. The mobilization process presented in more detail below focuses on the situation in South Africa in 1985–1989.

This application of the process model looks only at the ANC.  However, as mentioned earlier, other organizations attempting to mobilize the black population existed in South Africa.  Among the banned black opposition groups, AZAPO was the most prominent in the 1980s, appealing to a black-centric vision of a future South Africa (as opposed to the ANC's multiracial vision).  Although AZAPO was important as an outlet of an intellectual current within the black majority population, its strength was dwarfed by the ANC.

Among the tolerated black political groupings, the Inkatha Freedom Party (IFP) was the best known and most powerful.  The IFP was a populist ethnically based party, appealing to Zulu nationalism.  With a power base in KwaZulu (Natal), the IFP was a force that took the homelands policy's idea of self-government at its word.  The IFP's conservative, group-based (consociational) vision of South Africa, combined with its simultaneous opposition to *apartheid* and some animosity toward the Xhosa-dominated ANC, put the organization in the middle ground between the government and the ANC.  With the surge of ANC power in the mid-1980s, the IFP even cooperated with the *apartheid* regime against the ANC.  But the IFP's Zulu-centric outlook always constrained its appeal to non-Zulu blacks (indeed, IFP did not even succeed in attracting most Zulus, as there were more Zulus in the ANC than in IFP).  The IFP was an important regional actor in Natal and it could—and did—constrain the ANC in the latter's mobilization efforts there, but the IFP could not be a national-level actor. Even AZAPO exceeded the IFP in terms of potential appeal.

The focus of this section is on the ANC because of its importance, its wide appeal (in practice as well as potential), and its internationally recognized role as the main organization spearheading the effort to bring down the *apartheid* system. An analyst looking at South Africa in the late 1980s would have focused primarily on the ANC, since the organization had managed to mobilize successfully the black population in the mid-1980s. However, for the purposes of analyzing the mobilization appeal of other groups, a similar analysis of AZAPO and IFP might have sharpened the results and clarified why the other groups were minor actors in the main confrontation between the ANC and the *apartheid* state.

## Incipient Changes

There were at least four influential changes that, over time, shifted the demographic, economic, and political balance of power in 1980s South Africa.

**Demographic pressures.** Demographics always favored blacks in South Africa. But the different rates of natural increase among the various groups led to an ever-greater proportion of blacks to whites. The slowing of white emigration into South Africa to a trickle and declining fertility rates among whites contrasted with a high rate of natural increase among blacks. The first census in South Africa in 1911 showed that whites made up 22 percent of the population; the ratio remained similar—21 percent—in 1951, but it declined to 17 percent in 1970 and to 15 percent (including homelands) in 1980. The annual growth rate for blacks in South Africa in 1980–1989 was 2.39 percent. The corresponding figure for whites was 1.08 percent (1.82 percent for coloureds and 1.74 percent for Asians). To summarize, during the first 50 years of the country's existence, the share of whites had remained at approximately one-fifth of the overall population but then began to decline. Projections showed whites as making up a little over 10 percent of the population by 2000. The decreasing share of whites in the population and the corresponding increasing share of blacks exacerbated a whole range of economic and security problems for the government. Combined with the backpedaling on the homelands policy beginning in 1985, trends portended further changes in the direction of an ever-decreasing white share of the population.

**Decolonization and gradual fall of white minority regimes in southern Africa.** The *apartheid* policy was implemented during the era of colonial Africa (early 1950s). Despite the decolonization of the 1960s, southern Africa remained a bastion of white minority rule into the early 1970s. South Africa controlled Namibia (then South West Africa). The small states of Botswana, Lesotho, and Swaziland were black-ruled but economically dependent on South Africa and part of the South African Customs Union. Mozambique and Angola were Portuguese colonies. Zimbabwe (then Rhodesia) was ruled by a minority white regime. Even though sporadic guerrilla wars raged in Angola, Mozambique, South West Africa, and Rhodesia, larger states with a black majority rule were far away from South Africa (the nearest being Zambia). However, following the Portuguese coup in 1974, Angola and Mozambique became independent, ruled by black majority Marxist-leaning governments openly opposed to the South African government. Zimbabwe came under black majority rule in 1980. Namibia, enveloped in a guerrilla war, increasingly became a subject of UN attention. The formation of the Southern Africa Development Coordination Commission (SADCC), an attempt by the black majority states in the region to reduce economic dependence on South Africa, had the potential to reduce South African leverage over the neighboring countries.

The strategic changes confronted South Africa with a variety of unsettling developments, and the response was to overextend the country's armed forces and political posture. As guerrilla bases of an increasingly violence-prone ANC approached South Africa's borders, South African troops became involved in counterinsurgency operations in Namibia, conventional combat operations in southern Angola, and covert assistance to the antigovernment movements in Mozambique and Angola, all the while serving also as a tool to back up South African police in internal security operations. A nonaggression pact between Mozambique and South Africa in 1984 did not stop the ANC activities in South Africa. In 1988, South Africa agreed to withdraw from Namibia, losing another "buffer" state between the black majority states of Africa and the *apartheid* state. Thus, over the space of fifteen years (1974–1989), there had been a complete change in the strategic situation in southern Africa. By 1989, only South Africa remained under white minority rule, and the neighboring countries either sympathized with the ANC or supported it overtly. South Africa had lost its strategic glacis.

**Structural shift in South Africa's economy.** Until the 1970s, mining and agriculture had been the main sectors of the economy. The South African government accommodated both sectors with controls over labor and wages. The migratory labor system and homeland policy guaranteed a constant pool of labor, ensured government/industry control over wages, and kept unemployed blacks out of the area. In fact, the framework for the industrial residential areas, job reservation, and wage policies derived from the interests of the mining sector. Government management of the labor force in both instances enhanced the development of each industry by guaranteeing a supply of cheap labor.[14]

But the gradual shift toward a more developed capitalist economy led to the manufacturing sector dominating the economy, so that by 1973, the total value of manufacturing exceeded agriculture and mining combined. The manufacturing sector needed a more skilled and mobile work force, and the arrangements stemming from the domination of mining and agricultural interests became increasingly out of date. Parallel to the growth of manufacturing, black labor unions began to grow enormously in strength and determination, demanding higher wages and better working conditions. To maintain a stable manufacturing environment, businesses negotiated directly—and illegally—with unions. Thus, negotiations outside the *apartheid* framework and the rigid system imposed by industrial legislation became the norm.[15] The initial successful experiences with black trade unions spurred more businesses to negotiate and reach such settlements, providing an important indicator that the *apartheid* state was fracturing. In addition, business began to criti-

---

[14]"Decolonization in Southern Africa and the Labor Crisis in South Africa: Modernizing Migrant Labor Policies," in R. Hunt Davis (ed.), *Apartheid Unravels*, Gainesville: University of Florida Press, 1991.

[15]"The Black Trade Unions and Opposition Politics in South Africa," in Edmond J. Keller and Louis A. Picard (eds.), *South Africa in Southern Africa: Domestic Change and International Conflict*, Boulder, CO: Lynne Rienner Publishers, 1989.

cize the government's unwillingness to accommodate its needs for a different type of work force and greater flexibility in using it. The contradictions between the laws of capital and those of *apartheid* created a reluctant and distrustful partnership between labor and industry. In short, the structural change in the economy made *apartheid* laws seem inefficient and ill-adapted to contemporary economic needs, and they began to appear as more of a liability than an asset.

**Increasing international isolation and sanctions against South Africa.** Although South Africa had been subject to repeated censure by the United Nations, international sanctions against the country stayed in the realms of culture, sports, and diplomacy until the mid-1970s. Though a nuisance, the sanctions did not carry any great strength. But the sanctions gradually moved to the economic realm. In 1974, South Africa lost representation on the board of the IMF. Then, in 1977, the UN imposed a mandatory arms embargo on South Africa (a 1963 arms embargo had been "voluntary"). After violence broke out in South Africa in 1985, France terminated financial and economic relations, the British government banned new loans to the South African government and its agencies, and the United States imposed comprehensive sanctions on the country. The formal sanctions were accompanied by investor-led disinvestment and divestment pressures on U.S. corporations doing business in South Africa, resistance on the part of most foreign corporations to reinvest earnings, and refusals by banks to roll over loans. The sanctions began to hurt economically, weakened the regime's standing with the business elite, and brought home the realization to whites that the situation would only become worse, as South Africa assumed the role of an international pariah.

In addition, one major and unexpected event upset the balance of power in South Africa in the late 1980s.

**The fall of communism.** Mikhail Gorbachev's reforms and the unraveling of the Eastern bloc had a profound impact upon the political situation in South Africa. On the one hand, the regime lost a major trump card. With the delegitimization of communism and the end of the Cold War, the regime's portrayal of the ANC as a tool for a communist takeover of South Africa lost relevance. The reduced perceived threat from the left also diminished the support that South

Africa had enjoyed in the West because of its strategic location. On the other hand, the ANC lost its main source of support and military aid, dampening its prospects for a military success against SADF. Indeed, South African contacts with the USSR, begun during the negotiations over Namibia, were developing well, and there was potential for mutually profitable economic arrangements between the two countries (perhaps leading to a diamond cartel). In any event, both the ANC and the government faced a new situation after 1987–1988.

## Tipping Events

At least two main tipping events elicited and propelled black mobilization. These events led to the massive wave of unrest in South Africa and should have been clear as "tipping events" to an analyst looking at South Africa in 1989.

**The introduction of a new constitution.** After several years of discussions of constitutional changes, parliamentary (whites-only) approval in September 1983, and an all-white referendum in November 1983, the constitution entered into effect in September 1984. Even though the majority of the coloured and Asian populations boycotted the elections to their newly established two parliamentary chambers in August 1984 (only 16–18 percent turnout), the constitution at least recognized some political rights of the coloured and Asian populations. However, the constitution still did nothing to include the majority black population. As such, its entry into force amounted to a reaffirmation of *apartheid* and was a powerful statement of exclusion. Massive rioting and unrest in black townships followed. The unrest was accentuated by the focusing of the world's attention on South Africa after the award of the Nobel Peace Prize to Bishop Desmond Tutu (a black activist against *apartheid*) at about the same time.

**Botha's Rubicon speech.** Following the outbreak of serious violence and unrest in South Africa's townships, the government declared a state of emergency in 36 districts in July 1985. Amidst world attention to the unfolding of events in South Africa, the August 15, 1985 speech by South Africa's president, P.W. Botha, was widely anticipated as a potential breakthrough. Botha was expected to announce a fundamental concession to black demands, such as releasing

Nelson Mandela (the imprisoned leader of the ANC), completely eliminating the pass laws, declaring South Africa a unitary state, or opening a fourth chamber in the parliament for urban Africans. Instead, Botha delivered what was widely viewed as an affirmation of *apartheid*. In response, the ANC call upon its followers to make South Africa ungovernable was largely heeded, leading to prolonged unrest and rioting and a declaration of a nationwide state of emergency in June 1986 (renewed repeatedly until 1989).

## Leadership

Nelson Mandela was a leader with organizational talent and unparalleled moral authority among the black population of South Africa. Mandela started out as a high-ranking member of the African National Congress Youth League (ANCYL) in the 1940s and was involved in drawing up ANC's Program of Action to protest the passing of the *apartheid* laws in 1948. The program transformed the ANC (founded in 1912) from a moderate organization appealing to educated blacks into a mass political movement. Following the police action at Sharpeville in 1960, Nelson Mandela founded the ANC's military arm, *Umkhonto we Sizwe*. Having received military training in Algeria during that country's war of independence, Mandela proved an able military organizer. However, in 1963, the South African police discovered *Umkhonto*'s headquarters and captured Mandela and other ANC leaders.

In 1964, Mandela was sentenced to life imprisonment on a charge of sabotage. The decapitation of its leadership led to a virtual collapse of the ANC structure in South Africa, and the organization moved into exile. Although imprisoned, Mandela remained the spiritual embodiment of the black struggle for equal rights. His eloquent writings, smuggled out of prison, continued to influence the development of ANC strategy and thought. The fact that Oliver Tambo, an old colleague of Mandela dating back to the ANCYL days, came to assume leadership of the ANC only strengthened Mandela's resonance and influence.

Mandela's consistent emphasis on a multiethnic South Africa and sensitivity to the fears of black majority rule by many whites, combined with an accurate reading of the shifts in thinking among blacks in South Africa, even won him grudging respect among the less devoted *apartheid* supporters as well as more militant black elements. In short, Mandela emerged as a symbol of the ANC and opposition to *apartheid* in general. Although Mandela was not the tactical on-the-ground leader of the ANC, his views on the general conduct of the ANC's actions carried enormous weight and shaped ANC policy as well as black (and international) attitudes in general. His leadership skills were evident in his pattern of behavior: first, almost instinctively, Mandela expressed his own—often controversial—views and then, through his power of persuasion, hoped that others would agree.[16] Other leaders, such as Cyril Ramaphosa, skilled at organization and negotiation, provided capable leadership and guidance to the ANC and worked out its tactics.

## Resources and Organization

Given its largely impoverished base of support in South Africa and no personal wealth among its leadership, the ANC had few in-country resources. The ANC did not extract any monetary contributions from supporters in South Africa. Resources in support of ANC's activities came primarily from external sources.

What it lacked in resources, the ANC made up for in organization. The leadership of ANC consisted of a 30-member National Executive Committee (NEC), elected by a general congress of ANC's voting members and subject to approval by the group's president. A variety of constituent groups made up the ANC, including *Umkhonto we Sizwe*, ANCYL, guerrilla training camps, the Women's Section, officials from the ANC headquarters (in Lusaka, Zambia in the 1980s), and delegates from South Africa. In between its congresses, the executive body of the NEC was the National Working Committee (NWC), appointed by the NEC from within its ranks. Two coordinating bodies with executive powers steered activities between meetings of the NWC: the Politico-Military Council (PMC) and the External Coordinating Council (ECC). All departments fell under one of three offices: (1) the Office of the President (head and chief directing officer of the Congress and commander in chief of *Umkhonto we Sizwe*); (2) the Office of the Secretary General (chief administrative

---

[16]Nelson Mandela, *Long Walk to Freedom,* New York: Little, Brown, 1994.

officer of the movement); and (3) the Office of the Treasurer General (chief custodian of the funds and property of the ANC).[17]

The Political Committee had the task of mobilizing mass action in South Africa, establishing underground political and *Umkhonto we Sizwe* units, maintaining contact with legal organizations inside South Africa, setting up legal organizations within South Africa for mass mobilization and mass action, and reporting regularly to the PMC on the state of the internal organization and the conduct of political activities.  The Women's Section was specifically targeted to mobilize women outside and inside South Africa, organize international material and political support from women overseas, issue propaganda material for both internal and external use, and report regularly to the Secretary General.  ANCYL was focused on the recruitment of youth and students in South Africa into the ANC, organization of the ANC youth abroad into active units, and issuance of propaganda materials inside and outside South Africa.

The Military Headquarters directed the military operations inside and outside South Africa, subject to plans approved by the PMC/NWC/NEC, established underground *Umkhonto we Sizwe* units inside the country, and reported regularly to the NEC on the state of the army and the conduct of military operations.  Individual ANC guerrillas, after training in one of the camps in southern Africa (Zambia, Tanzania, Zimbabwe, Angola), were smuggled back into the country with forged papers and, using safehouses and sympathizers, contacted established ANC cells in the country and used weapons caches already in place.

The ANC had an extensive network of informal contacts with predominantly black organizations in South Africa, such as trade unions, student groups, and churches.  For example, the ANC provided leadership and organizational skills to the Federation of South African Trade Unions (FOSATU), which later became the Congress of South African Trade Unions (COSATU).

---

[17]"Report of the Commission on National Structures," *Constitutional Guidelines and Codes of Conduct Adopted at the Second National Consultative Conference of the African National Congress,* June 1985.

In the 1980s, then, the ANC possessed a highly sophisticated and far-reaching organizational structure. Its multitude of ties and contacts in South Africa meant good intelligence, while its sympathizers could be found in almost every nonwhite organization in South Africa and could shape political mobilization. The ANC's military wing also had a core of capable guerrillas and a well-developed network in South Africa.

## Foreign Element

The black majority-ruled states of Africa (and African states in general) provided diplomatic support and substantial assistance of their own and acted as a conduit for other foreign aid to the ANC. In terms of outright funding, Nigeria, Algeria, Egypt, Gabon, Cote d'Ivoire, and Senegal all contributed at least $1 million annually during the 1980s, but almost all African governments (with some exceptions, such as Malawi) provided some support. Other African states, such as Tanzania, Ethiopia, Zambia, Zimbabwe, and Angola provided training camps and facilities for use by the ANC. The Organization for African Unity (OAU) channeled some of the Soviet bloc aid to the ANC.

During the Cold War, the strategic location of South Africa and its strong anticommunist policy made it a strategic—if embarrassing—ally of the United States and NATO. Soviet support to the ANC and other black insurgent movements in southern Africa increased steadily with the rise of Soviet assertiveness in the developing world in the 1960s and 1970s, especially after the Soviet-supported Cuban military intervention in Angola. As South African troops clashed directly with Cuban forces and Soviet-trained and equipped Angolan troops, South Africa became the main Soviet adversary in Africa. The USSR and other Soviet bloc countries (primarily East Germany) were the main sources of military equipment and training to the ANC. The Soviet bloc also provided diplomatic support and funding. The USSR was probably the single most important source of funding and aid to the ANC in the early 1980s.

Disapproval of the *apartheid* policies led to widespread support for the ANC from West European and North American sources. A variety of social-democratic organizations in Western Europe (such as the West German Social-Democratic party) and governments ruled by social-democratic parties (especially the Scandinavian countries) contributed funds to the ANC. Through a variety of contacts with political parties, religious organizations, student groups, and trade unions, the ANC received support from most of the developed Western countries, including the United States. The UN contributed to funding for the ANC through its financial support for the ANC office at the UN.

In short, there was widespread international support for the ANC. Some of it was tied to the Cold War and Soviet use of black majority rule organizations for its own ends. But support also came from groups in developed Western countries sympathetic to the anti-*apartheid* cause. In 1986, the ANC claimed that more than half of its aid came from non-Soviet sources, though that may have been an exaggeration.

## Overall Assessment of Mobilization

The gradual transformation of the South African economy increased the leverage of blacks with the white business elite, while demographic pressures, international isolation, and growing vulnerability of South Africa to penetration by armed ANC groups placed the South African government on the defensive by the early 1980s. After Botha's unexpected strong defense of the *apartheid* system and his seeming challenge to the opposition, the ANC's strong appeal to the blacks of South Africa, its moral leadership under Nelson Mandela, and its worldwide support, combined with its sophisticated organization and a coordinated plan of action, made the organization well able to implement its call to "make South Africa ungovernable" in the mid-1980s. Other anti-*apartheid* black groups existed in South Africa, but none could match the standing and the clear ability of the ANC to mobilize its supporters.

## ASSESSING THE STATE

### Accommodative Capability

How inclusive and responsive were South African political structures to the general populace? A relatively open political system and regular elections served as channels of elite replacement, but the mechanisms extended primarily to whites and secondarily to the coloureds and Asians.  The 1984 constitution amounted to a far-reaching step by the whites toward power sharing with the coloureds and Asians.  Though the political power of the two groups remained more nominal than real, their political rights were increased and they obtained a representative voice in the parliament.  However, the 70 percent of the population classified as blacks remained without any political rights.  These restrictions were written in the constitution.  Given such a state of affairs, in an overall sense the political structures were highly exclusive.  The political system amounted to a race-based oligarchy, with accountability extended only to the whites.

What potential for change in political structures existed in South Africa in the late 1980s?  The constitution established a strong Presidency, giving it a virtually free mandate to decide national issues with only limited oversight (primarily indirect) by the House of Assembly.  Since the Afrikaner-dominated NP controlled both the House of Assembly and the Presidency, and since their specific understanding of identity gave Afrikaners strong in-group solidarity, for all practical purposes the political structures could not change peacefully without the approval of the Afrikaners.

The political system did not provide for any meaningful interest-articulation mechanisms for blacks.  Instead, the political system tried to establish separate entities that would provide limited (primarily local) self-government and political representation for blacks outside the main political system in South Africa.  Without any political representation of the interests of blacks at the national level, the system simply was not equipped to handle conflict resolution between blacks and the state.  Since the constitution precluded the formation of any national-level black representation, the potential for fundamental change in political structures was low.

In short, in terms of inclusiveness, the South African political system was explicitly exclusive.  The 1984 constitution allowed for the nominal inclusion of the two nonwhite groups but did nothing to address the country's black majority population.  The system had limited flexibility in that it served primarily as a tool of Afrikaners.  If a white majority (based on the Anglophones and at least a strong minority, if not the majority, of Afrikaners) decided that the system had to change, they could implement fundamental reforms.  But such a transformation depended on a change of views toward *apartheid* among the Afrikaners. The best that can be said is that the South African political system had some built-in flexibility within the oligarchic structure.

Prevailing norms of governance varied according to racial classifications. Among the whites there was a good deal of tolerance of differences, as long as the opposition did not step outside the bounds and engage in outright illegal activities. Even then, state sanctions against such individuals were relatively mild. The system showed less tolerance toward the coloureds and Asians and showed very little tolerance toward any displays of opposition by blacks. The premise of the state was that blacks were not fit for modern political institutions and instead were better suited for traditional ("tribal" chief) modes of governance. The state saw actions by blacks aiming to change South Africa's political system to be essentially impermissible, and even nonviolent acts of disobedience by blacks could be— and sometimes were—treated as fundamental threats to the state.

In the South African polity, racially based collectivist norms were clearly superior to individual rights and underpinned the whole structure of governance. Such norms stemmed from the fundamental racist outlooks toward the various groups, a racially based hierarchy, and different views as to the "proper" role of specific groups in the society. The views had an ideological—indeed, theological— explanation in the *apartheid* doctrine and seemed deeply internalized, especially among the Afrikaners.

What was the level of cohesion among the ruling elites? For almost two decades after the NP's victory in 1948, the party was dedicated to a radical form of Afrikaner nationalism and united in its uncompromising attitude about implementing *apartheid* and in its resentment of Anglophones in South Africa. Differences over the interpretation of *apartheid* policies in the 1960s (such as allowing visiting black diplomats to stay in white hotels) provoked a split within the NP into two wings, the *verkrampte* (uncompromising) and *verligte* ("enlightened" or pragmatic), as well as the formation of an extreme-right splitist party, the *Herstigte Nasionale Party* (Reconstituted National Party). By 1982, the *verligte* wing's discussion of accommodating some urban blacks in the South African political system led to another split and the formation of the extreme-right Conservative Party (CP) by a group of deputies expelled from the NP. The reforms launched by Botha in the mid-1980s led to further splintering within the NP, with the *verkrampte* wing openly blaming the reforms and the liberal wing of the NP for the outbreak of violence. In addition, prominent members of the NP (Denis Worrall and Wynand Malan) left the NP and formed political parties dedicated to faster political reform.

In the parliamentary elections in May 1987, the NP kept its majority. The main political party of the Anglophones, the Progressive Federal Party (PFP), lost ground, as many Anglophones voted instead for the NP. The Conservative Party made large gains and became the main opposition party in the parliament. Several additional extreme-right Afrikaner groups (on the neo-fascist fringe) also formed.[18] Under such conditions, the NP, led by the *verligte* wing, emerged as an increasingly centrist organization.

By early 1989, Afrikaners were no longer politically united. As early as 1987, the *Broederbond* (Brotherhood),[19] a highly influential secret Afrikaner society, had accepted the fact that black participation in South African politics at the national level was inevitable and only a matter of time. The militant Afrikaners had bolted the NP for the far-right Afrikaner groups. An alliance of four moderate and liberal parties (two Anglophone and two Afrikaner), aiming to dismantle *apartheid,* formed a united opposition party, the Democratic Party. Even some leaders of the Dutch Reformed churches declared *apartheid* evil and a sin against God.

Clearly, Afrikaners had lost their unity. Neither upholding strict *apartheid* nor anti-Anglophone proclivities remained as pillars of the NP. From a mechanism of militant and segregationist Afrikaners, the NP was evolving into a vehicle for moderate whites (Anglophones and Afrikaners).[20] A substantial portion of the Afrikaners turned

---

[18]Adrian Guelke, "The Quiet Dog: The Extreme Right and the South African Transition," in Peter H. Merkl and Leonard Weinberg (eds.), *The Revival of Right-Wing Extremism in the Nineties,* London and Portland, OR: Frank Cass, 1997, pp. 254–270.

[19]Membership in the *Broederbond* was by invitation only, after an Afrikaner had demonstrated a high level of loyalty and leadership to the Afrikaner cause. Almost all prominent Afrikaner political and business leaders (including President Botha) were members of the *Broederbond*.

[20]Anthony W. Marx, "Apartheid's End: South Africa's Transition from Racial Domination," *Ethnic and Racial Studies,* Vol. 20, No. 3, July 1997, pp. 475–496.

toward far-right political groups uncompromising in their stance toward *apartheid*. Still others rejected *apartheid* altogether. The lack of cohesion among Afrikaners was matched by an increasing unity among the whites in general about the need to move beyond *apartheid*.

In conclusion, the accommodative capability of the South African state defies easy labels. Democratic political institutions existed but were limited to certain groups, primarily the whites. If a fundamental shift in views among the whites occurred, the system had enough flexibility to address even the most basic questions. But this responsiveness was only extended to changing views among the privileged minority. Moreover, that minority had profited from the system and had many incentives to keep it in place. Norms of governance and a collectivist racially based view of the society indicated little tolerance of any black political activism or black aspirations to the national political level.

Still, there were numerous signs in place by early 1989 of a fundamental political realignment among the whites in South Africa, with a diminution of earlier Anglophone/Afrikaner differences and the emergence of a consensus among the whites that *apartheid* had to be dismantled. Black mobilization and consequent violence seems to have played a role in shaping the political realignment.

## Fiscal and Economic Capability

The fiscal health of South Africa in the 1980s was essentially sound, though there were some worrisome trends in evidence. Until 1986, South Africa had a high rating with international lending institutions, a strong history of debt repayments, and a manageable debt. But the various international sanctions and especially the outflow of investment (the book value of U.S. investments in South Africa fell from $2.4 billion in 1982 to $1.3 billion in 1986) and the stopping of loans to South Africa led to current account problems. A capital flight of $1 billion occurred in the six months between September 1985 and March 1986. The South African currency (the rand) plummeted from US$1.28 to US$0.39 in 1986, where it stabilized until 1989. As Table 4.10 illustrates, the economy continued to grow, though at a sluggish pace, inflation reached worrisome double digits, and the government engaged in sustained deficit spending (though not at alarming levels)

**Table 4.10**

**South African Revenues and Expenditures (Millions of rand)**

|                                | 1985–1986 | 1986–1987 | 1987–1988 | 1988–1989 |
|--------------------------------|-----------|-----------|-----------|-----------|
| Total revenue                  | 29,851    | 34,611    | 38,794    | 51,234    |
| Total expenditure              | 33,026    | 40,213    | 46,319    | 59,923    |
| Deficit                        | –3,175    | –5,602    | –7,525    | –8,689    |
| Deficit as percent of revenues | 10.6%     | 16.2%     | 19.4%     | 17.0%     |

SOURCE: *The Statesman's Yearbook,* 1988–89 and 1990–91.

in the late 1980s.  In addition, South Africa largely lost the ability to obtain new loans due to international sanctions.

The worsening economic situation notwithstanding, it would be an exaggeration to say that South Africa faced fundamental economic problems.  As the world's leading producer of gold (about 30 percent of world production) and a major producer of diamonds and a variety of rare metals, South Africa was assured of a steady income. However, the international isolation and sanctions had hurt the country's manufacturing and services industries and threatened to cause further disruptions.

As for the structure of the state budget in the late 1980s, defense already occupied a prominent position among the main budgetary categories, as shown in Table 4.11.  Since expenditures on social services for nonwhites were already at low levels, further outlays on defense would have been taken at the expense of social spending for whites.

As Table 4.12 shows, the majority of South African government revenues came from income taxes and of those, most came from income taxes on individuals.  Under the South African tax structure in the late 1980s, a progressive tax rate was applied to incomes ranging from 16 percent on the first R10,000 to 53.5 percent on income exceeding R42,000.  Because of the concentration of wealth, the main tax burden fell upon the whites, as Table 4.13 indicates.  Although the *apartheid* system served the whites well, it was not cheap.

South Africa had an effective machinery for extracting more resources from the society.  However, the portion of the society that

## Table 4.11

### Structure of South Africa's Budget, 1987–1988 (Millions of rand)

|                        | 1987–1988 | Percent of State Budget |
|------------------------|-----------|-------------------------|
| Defense                | 7,018     | 15.2                    |
| Economic services      | 6,248     | 13.5                    |
| Education              | 8,617     | 18.6                    |
| Health                 | 4,339     | 9.4                     |
| Other social services  | 4,274     | 9.2                     |
| Interest on public debt| 6,098     | 13.2                    |
| Other                  | 9,725     | 21.0                    |
| Total                  | 46,319    | 100.0                   |

Percentages may not total to 100 due to rounding.
SOURCE: *The Statesman's Yearbook,* 1990–91.

## Table 4.12

### South Africa's Revenue Sources, 1987–1988 (Millions of rand)

|                   | 1987–1988 | Percent of Total Revenues |
|-------------------|-----------|---------------------------|
| Income tax        | 21,887    | 56.4                      |
| General sales tax  | 9,954     | 25.7                      |
| Excise duties     | 1,920     | 4.9                       |
| Customs duties    | 1,540     | 4.0                       |
| Other             | 3,493     | 9.0                       |
| Total             | 38,794    | 100.0                     |

SOURCE: *The Statesman's Yearbook,* 1990–91.

## Table 4.13

### Income Tax Payable by Individuals, South Africa, 1986

| Group    | Average Income (rand) | Tax, Not Married (rand) | Tax, Married with Four Children (rand) |
|----------|-----------------------|-------------------------|----------------------------------------|
| Black    | 5,016                 | 123                     | 0                                      |
| Coloured | 6,732                 | 363                     | 0                                      |
| Asian    | 9,240                 | 849                     | 82                                     |
| White    | 18,732                | 3,361                   | 2,265                                  |

SOURCE: *South African Yearbook,* 1986.

formed the state's core support group was already taxed heavily. Especially at the upper income levels, tax rates were unattractive in comparison with some of the other developed states. Conceivably, the government could have increased rates further, but such a move threatened to be counterproductive in terms of causing an outflow of capital.

The core support group of the *apartheid* state, the whites, differed in terms of their income and political party support. Surveys of party supporters in the late 1980s showed a pattern of the wealthier whites supporting the political party most identified with the goal of dismantling *apartheid,* while the lower-class whites supported the far-right parties. The pattern indicated in Table 4.14 of the poorer whites especially adamant about protecting their social standing and income against potential competition from blacks is not unusual. Nor is the pattern due to the different levels of wealth between Afrikaners and Anglophones since, by the 1980s, income levels between the two groups had become similar.

In terms of the NP's overall dependence on support from the whites, the sympathies of the poorer whites with the far-right Afrikaner parties made political appeals to that stratum of the population not all that enticing, compared to an attempt to coopt the wealthier, more liberally inclined whites. The latter approach allowed for the possibility of increased financial support for the party, though it meant liberalizing and perhaps dismantling *apartheid*.

Table 4.14

Household Income and Political Party Support Among Whites, South Africa, 1987

| Household Income | PFP (liberal) | NP (centrist-right) | HNP and CP (far-right) |
|---|---|---|---|
| Upper-middle | 38% | 20% | 10% |
| Middle | 43% | 45% | 49% |
| Lower-middle | 19% | 35% | 41% |
| Sample size | N = 335 | N = 801 | N = 262 |

SOURCE: Peter Berger and Bobby Godsell (eds.), *A Future South Africa: Visions, Strategies, and Realities*, Boulder, CO: Westview Press, 1988, p. 28.

In conclusion, the fiscal and economic capability of the South African state in the late 1980s defies easy labels. Economic problems certainly existed, caused to no small extent by international sanctions, and worrisome trends were in place (deficit spending, slow growth, and inflation). However, the economy remained essentially sound and the country's unique assets guaranteed that economic problems would not become catastrophic. In terms of the state's ability to extract further resources, the machinery to do so existed and the core support group remained wealthy, even if already taxed at significant levels. The more interesting development was the political splintering among the whites—in their views of *apartheid*—more along income levels rather than Afrikaner/Anglophone distinctions. The wealthier whites, whether Anglophone or Afrikaner, increasingly questioned the wisdom of continuing *apartheid*. And yet, it was the wealthy whites to whom the ruling elite had to appeal and take into account.

## Coercive Capability

There was no ambiguity about the authorities' access to the command and control channels of the apparati of violence nor about the propensity to use them against domestic opponents. The 1984 constitution established a strong presidency and named the president the commander-in-chief. Under Botha's tenure, the State Security Council (SSC) was revitalized as a crucial committee at the Cabinet level charged with security.

The SSC secretariat included several directorates: strategy, national intelligence, strategic communication, and administration. The SSC secretariat was also in charge of interdepartmental committees, which oversaw the implementation of strategy at the bureaucratic level, and joint management centers, which ensured the extension of the strategy to the local level. The whole system, known as the National Security Management System (NSMS), amounted to an elaborate security executive mechanism, outside the bounds of any parliamentary supervision.

The minister of defense had direct civilian supervision over the military. The chief of the SADF had the operational command of the armed forces and was accountable to the minister of defense for carrying out the government's policy for the armed forces. The SADF consisted of three services (Army, Navy, Air Force). Chiefs of each service reported directly to the SADF Chief. Besides being responsible for defending South Africa's borders, the Army had the task of preserving state authority and assisting the South African Police (SAP) in internal security operations. The Air Force, through its mission of support to the Army, also had a specific internal security role. The Ministry of Law and Order controlled the SAP.

The legal mechanisms justifying the use of force domestically stemmed from a number of laws passed primarily in the 1950s and 1960s against the opposition Congress Alliance (composed of the ANC, Indian Congress, and the South African Communist Party). The laws included: (1) the Internal Security (Suppression of Communism) Act, 1950 (gave the government the power to ban a wide range of political activities and organizations); (2) the Public Safety Act, 1953 (allowed for the declaration and regulation of a state of emergency); (3) the General Law Amendment Act, 1962 (made sabotage a capital offense); (4) the General Law Amendment Act, 1963 (legalized detention without trial for up to 90 days); (5) the Criminal Procedure Amendment Act, 1965 (extended detention without trial for up to 180 days); (6) the Terrorism Act, 1967 (introduced indefinite detention of persons suspected of terrorism); and (7) the Internal Security Act, 1982 (consolidated all of the above security legislation under the ordinary law).

Who served in the apparati of violence? The military and police apparati were heavily staffed by Afrikaners. There is little doubt that the personnel in the two organizations could be relied upon in case of domestic unrest.

The SADF was divided into the Permanent Force, the Citizen Force, the Commandos, Auxiliary Services, and civilians. Professional soldiers made up the Permanent Force and were responsible for the training and administration of the entire force. The Citizen Force was composed of noncareer soldiers and supplemented the Permanent Force. The Commandos functioned primarily as paramilitary home defense (area protection) units. In 1989, the active personnel strength of SADF was 106,400 (including 64,100 conscripts), some 175,000 active reserves, and 130,000 Commandos. Conscription was mandatory for white males (voluntary for others) at age 18, and the term of service lasted two years. Following the term of service, conscripts then had to serve two years over the next 12 years. Following that term, they qualified for the Active Citizen Force Reserve. After five years in the Reserve, they could be called up and assigned to the Commando forces.

Nonwhite volunteers served either as career personnel in the Permanent Force or on a temporary basis in the Citizen Force or Commandos. The nonwhites were assigned to separate units (often specialized). Virtually all members of the Permanent Force were whites and almost 99 percent of the Citizen Force was composed of whites. In 1987, whites made up approximately 85 percent of the Commandos (blacks made up 4 percent, coloured 9 percent, and Asians 2 percent).[21]

The officer corps of the SADF was overwhelmingly Afrikaner. In its symbols and mythology, the SADF had strong Afrikaner tendencies (for example, one of its main training institutions was the Danie Theron Combat School, named for an Afrikaner with anti-British proclivities). Generally, the higher the rank, the greater was the percentage of Afrikaners and the fewer of Anglophones.

The South African Police had an active strength of 60,950 (32,754 whites and 28,196 nonwhites) in 1987, supplemented by some 20,000 reserves. Generally, black police were not allowed to carry arms. The high command of the SAP seems to have been primarily composed of Afrikaners.

What norms existed with respect to the use of violence domestically? The state had used force (police as well as the military) repeatedly against domestic opponents since the inception of the *apartheid* system. On several occasions (Soweto in 1976, mid-1980s in many urban areas), the death toll among the protesters ranged into the hundreds. The internal security apparatus also had a reputation for excessive brutality (Sharpeville, 1960) and was implicated in instances of torture and murder of prisoners (Steven Biko, 1977) and detainees (in 1984–1986, 18 people supposedly committed suicide

---

[21]*Survey of Race Relations—1987*, Johannesburg: South African Institute of Race Relations, 1987.

while in detention).[22]  The internal security intelligence apparatus engaged in assassination of suspected opponents, often with the use of proxies (KwaMakutha raid, 1987).  While some of the targets of the internal security apparatus were ANC guerrillas and many operations had a counterinsurgency or violent riot-control character, at least some of the actions of the security services fell more in the realm of intimidation of opponents and use of excessive force.

The South African justice system did not hesitate to impose death sentences for incidents related to urban unrest.  For example, in the period 1985–1988, 101 people charged with such offenses were sentenced to death (17 were executed).  In short, the South African internal security apparatus had a reputation for dealing harshly with nonwhite, and especially black, opponents.  Officially, the government considered South Africa under siege, and it treated the black unrest and ANC-sponsored sabotage as a foreign-inspired communist guerrilla war.

Was the force suitable for domestic use?  One of the missions of the SADF was internal security, and its equipment and training reflected preparation for the task.  Many of the armored personnel carriers used by the Army were suitable for domestic internal security duties.  The Air Force used several squadrons of Impala aircraft for counterinsurgency duties.  Nonetheless, the military was never at ease with its domestic role and, when used, played a backup role to the police.  The primary tools of the state in quelling domestic unrest were the Security Police and the SAP, a heavy police force equipped with substantial riot gear and armored vehicles for patrols in areas where government control was weak.

The internal security apparatus had extensive domestic intelligence capabilities.  Three intelligence services were engaged in domestic intelligence collection, coordination, and analysis:  the Security Police (domestic security intelligence collection and counterintelligence), the National Intelligence Service (or NIS, intelligence processing and coordination), and the Directorate of Military Intelli-

---

[22]The report of the post-*apartheid* South African Truth and Reconciliation Commission provides proof of the police involvement in the death of Biko and other prisoners. See *Truth and Reconciliation Commission of South Africa Report*, Capetown: Juta, 1998.

gence (or DMI, military and defense intelligence inside and outside South Africa). The intelligence organizations engaged in covert operations against the ANC and its suspected sympathizers.

In conclusion, the coercive capability of the South African state was strong. The state had arrayed a powerful force against internal opponents and showed the determination and will to use the means of coercion on numerous occasions. The top executive authorities had a virtually free hand in ordering force to quell unrest and had set up an elaborate command and control mechanism for effective coordination of all internal security activities. Afrikaners were prominent in the apparatus of coercion and dominant at its higher echelons.

## STRATEGIC BARGAINING

This analysis of the mobilization of the ANC and the capabilities of the South African state to deal with such a mobilization provides a second example of how the interaction between a mobilized group and a state can be thought of within the framework presented in Chapter Two. Similar to the analysis of Yugoslavia presented in Chapter Three, the group and state types are categorized on the basis of their capacities, and the matrices provide a way to think about their interaction conceptually.

### Measuring the Group's Capacities

Concerning the leadership of the ANC-mobilized blacks, Nelson Mandela was as devoted to the ANC and majority rule in 1989 as he had been in 1960 or in 1948, despite spending 25 years in a harsh prison. His moral stature and wide appeal had only increased.[23] A core of skilled organizers directed the ANC and provided capable leadership in Mandela's absence. All of the top leadership of the ANC were convinced of the justice of their cause and unwaveringly dedicated to it. The ANC leadership's acceptance of violence and their call in the mid-1980s to make South Africa ungovernable shows

---

[23]Robben Island, the maximum security prison where he was held, was known among anti-*apartheid* activists as Mandela University, because of the political mentoring and leadership he provided to the movement.

that they were willing to take risks. It was by no means assured that the black population would heed the ANC call for massive unrest. Had the call gone largely unheeded (or had it been deterred by a show of force by the South African internal security apparatus), the ANC would have suffered a major blow to its prestige, for its claim to have wide resonance among the black population would have been exposed as false. Thus, the assessment of leadership is "strong."

Concerning the ANC's resources and organization, previous observations paint a mixed picture. On the one hand, the ANC had tremendous worldwide sympathy, enough to cause South Africa to suffer international isolation and be treated as a pariah country. On the other hand, the ANC had no domestic resource base to speak of and depended on sometimes fickle financial support from foreign sources. And compared to the South African state, its resources were puny. The fall of communism and the slowing of Soviet aid to the ANC was troublesome. Moreover, the Soviet military support the ANC had received was well suited to forcing the state to spend enormous amounts on defense. However, the resource base was sufficient to meet all near-term goals, such as making South Africa ungovernable, and many mid- and even long-term goals (bringing the South African state to its knees through overextension and isolation). The ANC also showed that it used the resources effectively. Support for the ANC had increased greatly in the 1980s, and there were good prospects that the level of support would at least remain steady. Western support was unlikely to be in the military realm (and would not replace the Soviet bloc support), though other forms of aid too may have been effective. Thus, the assessment of resource support is "good."

Regarding popular support for the ANC mobilization, there is little question that the ANC had the support of the majority of the black population of South Africa. Neither the black nationalist organizations that competed with the ANC for support among blacks nor the group-based organizations (such as IFP) ever achieved anywhere near the influence of the ANC. And whereas the ANC could subsume the other groups under its umbrella, the other groups could not do the same with the ANC. In the same vein, the multiracial unionism represented by the ANC also elicited support from segments of all other groups in South Africa (coloureds, Asians, and whites). Thus, the assessment of popular support is "broad."

Based on these assessments, the blacks mobilized by ANC are a type A group. The capacities of such a group are as follows:

Accommodative: high;

Sustainment: high;

Cohesiveness: high.

Note that if resource support is assessed as weak, then the group type is C, with accommodative and cohesiveness capacity remaining high and sustainment changing to low.

## Measuring the State's Capacities

Concerning the leadership of the state, all of the observations compiled paint a picture of the Afrikaner leadership in crisis and a political realignment under way by early 1989.  Just as Botha was embroiled in a power struggle within the NP (F.W. de Klerk replaced Botha as the NP leader in February 1989), the Afrikaner elements unreconciled to any easing of *apartheid* had become the main opposition to the NP in the parliament.  Moreover, the NP had lost its full identification with Afrikaners and was increasing its support among Anglophones.  Finally, the heretofore weak liberal white parties had united.  Under such circumstances, the Afrikaner leadership (whether moderate or extreme right wing) was conscious of other leaders and engaged in intense political maneuvering.  Any leader appealing to a narrow Afrikaner political spectrum was certain to become marginalized in the new intrawhite political environment. Since the Afrikaner political unity had splintered and the Anglophone/Afrikaner division no longer had as great a resonance, making appeals to the greater polity of whites was the only solution for a leader with truly national aspirations.  Finally, in conditions of flux and realignment, old political rules and certainties no longer applied, and dedication to the goals of a previous era was not a wise policy for political leaders.  Thus, the assessment of leadership is "weak."

There is some ambiguity about the state's fiscal capacity.  On the one hand, the state had engaged in substantial deficit spending, defense already occupied a prominent place in the budget, and the core group supporting the state, the whites, already faced hefty taxes.  On the other hand, deficit spending had not been that lengthy or deep, there was room for some reallocation of funds, and the state probably could have extracted more resources from the whites had it chosen to do so. Moreover, South Africa remained the world's single largest producer of gold, it had almost unrivaled other mineral resources, and its economy continued to grow even during the years of unrest in the mid-1980s. In short, the state faced some fiscal problems but not a crisis, and its economy remained essentially sound. Thus, the assessment of fiscal position is "strong."

Regarding the regime type of the state, for the whites, the South African state was inclusive. Within that group, competitive elections took place, media were uncensored (except for specific instances concerning information about ongoing unrest) and there was an acceptance of democratic practices and tolerance of different views. The integration of coloureds and Asians into national politics, even if more nominal than real, was a definite step toward further inclusiveness. The elimination of a host of *apartheid* restrictions on blacks by the late 1980s also foreshadowed a potential further move toward political inclusion of the black majority of the population. Despite all of the above, however, the state remained an oligarchy, it denied almost three-quarters of the population any political rights, and it even stripped many blacks of citizenship. Thus, the assessment of regime type is "exclusive."

Based on these assessments, the South African state is a type H state. The capacities of such a state are as follows:

Accommodative: low;

Sustainment: high;

Coercive: high.

## Outcome of Bargaining and Preferences for Violence

Based on the matrix showing the preferences of the mobilized group, a type A group has the following preferences toward a type H state: (1) exploit, (2) intimidate, and (3) negotiate. Based on the matrix showing the preferences of the state toward a mobilized group, a type H state has the following preferences toward a type A group: (1) exploit, (2) repress, and (3) negotiate. Note that if the group type is C, due to an assessment of resource support as weak, the group's preferences stay the same: exploit, intimidate, negotiate. The state's preferences change to: repress, exploit, negotiate.

Comparing group and state preferences leads to the conclusion that the potential for violence certainly existed in the dyadic encounter between the ANC-mobilized black population and the *apartheid* regime. However, the more striking observation is that, despite a virtual state of war between the ANC and the state, neither side preferred to resort to violence. Instead, the preferred strategy for both sides was to compete with the other through largely nonviolent means. The preference sets of both sides were the same, namely, a strategy of exploitation designed to weaken the opponent, with a hedging strategy of using violence. Neither side trusted the other, but each respected the other's capabilities and held out the potential for a negotiated solution.

What is telling about the choice of strategies is the underlying sense of an impasse. The ANC's resorting to a full-blown strategy of violence would have been risky and would certainly have been costly. But the ANC was left with few alternatives other than to pursue its goals as forcefully as possible without crossing over the threshold of a mass uprising. The state was in a similar position. It commanded considerable coercive resources but preferred to use them as a deterrent and a tool to force the ANC to negotiate a transfer of power on the state's terms or, even better, to share power without the state losing control.

The matchup of state type H and group type A is a notable dyad in that the preference sets for both provide an interesting combination of least "bad" choices within a highly dangerous (violence-prone) set of dyads. The dangerous elements come into play because an exclusive regime leads to a state type that is highly prone to violence. However, among the four state types having an exclusive regime, type H is the least prone to violence. Also interesting, a type A group has an initial strategy of preference for violence against two of the exclusive regime state types, but not toward type H. In terms of the ANC-South African *apartheid* state matchup, the implication is that the specific capacities proved fortunate for the purposes of avoiding violence. For example, if the state had strong leadership and had been classified as type D, its first preference for a strategy against group type A would have been "repress." The South African state in the late 1970s and early 1980s was probably a type D state and, as the record shows, at that time the state showed no hesitation in using violence at home and abroad against the ANC.

In the alternative classification of the group, that is, if the group is assessed as having weak resource support and coded as type C, then the preferred state strategy changes to repression and the use of violence, with a hedging strategy of exploitation. In other words, the top two state preferences are the inverse of what they would be if the group were assessed as type A. The rationale for the change is straightforward: the state has low accommodative capacity and the group's low sustainment capacity compares unfavorably with the high sustainment capacity of the state.

## The Course of Events

The strategic choices outlined above provide insights into the way that events unfolded. The realignment in South African leadership became clear in 1990. F.W. de Klerk replaced Botha as president in August 1989, and the NP, under his leadership for several months already, won almost half the votes in the early parliamentary elections in September 1989. The extreme-right Conservative Party won almost a third of the votes, while the unified anti-*apartheid* Democratic Party won a fifth of the votes. F.W. de Klerk came from a different background and generation than Botha and did not have Botha's history of political run-ins with the Anglophones. The NP under de Klerk was a far different party from the NP of an earlier era. More of a white centrist organization, it ceased to be a tool of militant Afrikaners, and as the 1989 elections showed, it appealed to the moderate elements among Afrikaners and Anglophones united in the view that *apartheid* had to be dismantled.

Within eight months of the election, the government legalized the ANC and a host of other groups (including the South African communist party, now somewhat ideologically adrift after the fall of communism) and released unconditionally over one hundred political prisoners, including Nelson Mandela. In addition, the government lifted many of the emergency measures and, in May 1990, began direct talks with the ANC about power-sharing and a new constitution. In early 1991, the government abolished the Population Registration Act (which decreed the registration at birth of all people in South Africa by race), effectively putting an end to *apartheid* once and for all.  The convening at the end of 1991 of a multiparty conference to negotiate South Africa's transition to majority rule (the Convention for a Democratic South Africa, or CODESA) signified the beginning of the process of the actual transfer of power.  Following astute political maneuvering by de Klerk, in which the NP defeated the white elements unreconciled to the end of *apartheid,* and after protracted negotiations, the NP government and the ANC (though not some of the other groups involved in the negotiations) agreed on a provisional constitution for a post-*apartheid* South Africa.  The actual transition of power came in April 1994, with elections on the basis of universal suffrage.

In terms of the model framework, the state had been a type D in the 1970s and evolved into type H in the late 1980s as a result of shifts in the basis of support for the ruling elite.  The preferred strategy of a type D state was "repress," and then, as a type H, the preferred strategy changed to "exploit."  Moreover, the state became increasingly dissatisfied with a hedging strategy of "repress" and was moving toward the secondary hedging strategy of "negotiate."  Despite the abandonment of the "repress" strategy, the "exploit" strategy did not mean nonconfrontational negotiations.  As events showed, negotiations on transfer of power were suspended on several occasions, the state used the specter of intrablack violence and extreme-right white nationalism to pressure the ANC into making concessions and, when it began, the success of the whole process was far from a given.  But the "exploit" strategy did succeed in the state effecting in a nonviolent manner a transfer of power while ensuring group-level guarantees for the whites.

From the perspective of the ANC, the group's strategy of "exploit" and "intimidate" was paying off with the growing proclivity of the state to abandon its "repress" strategy as a hedging option and move toward a "negotiate" strategy.  However, the group faced the real problem of potentially losing public support as a result of intrablack violence.  (Organized violence was most prominent in Natal between the Zulu-dominated Inkatha Movement and the Xhosa-dominated ANC, but there were other instances of violence, for instance between the ANC and the more militant black republican organizations.)  If, over the space of protracted negotiations and loss of public support, the group type actually moved toward type D, then its ability to negotiate successfully with the state would have been damaged. A type H state would have reverted to a "repress" strategy against the ANC if the latter became a type D group. And a type D group would have lost much of its negotiating leverage against the state, for its threat of violence would have lost credibility and it would be forced to negotiate with the state from a position of weakness. In terms of the model, the change is illustrated by the top strategy preference changing from "exploit" to "negotiate," with "intimidate" relegated to a tertiary position. Just as with the state's "exploit" strategy, the group's "exploit" strategy did not mean nonconfrontational negotiations. The ANC issued an ultimatum to the state in 1991 and used a mass protest campaign in 1992 to exert pressure during the negotiations and political maneuvering. But the ANC's strategy was largely free of violence, especially against the whites, and it did succeed in sweeping aside other groups aspiring to leadership of the blacks and in preparing a dominant position for itself in post-*apartheid* South Africa.

Had an analyst in early 1989 used a framework similar to the one presented here to examine the situation in South Africa, the following intelligence needs would have become apparent:

- Attitudes and preferences toward political parties among the Anglophones and Afrikaners and the relationship between income and political party support;

- The extent of contacts and ties between SADF and SAP and the extreme-right Afrikaner groups;

- The impact of international sanctions on the South African economy and the perception of the impact by South African business elite;

- Sketches of alternative post-*apartheid* futures among the white political elite;

- The effect of the fall of communism on the ANC's resource base and on its views of post-*apartheid* South Africa;

- The flexibility and openness of the ANC to agree to preconditions (especially guarantees for whites) to negotiations with the state;

- The potential appeal of black republican groups and their strategies of action;

- The views of leaderships of the "homelands" toward the ANC and the extent of the state's leverage over them.

Close attention to these issues may have allowed for a better understanding of the trends and perhaps more effective U.S. support for the accommodation in South Africa.

## FINAL OBSERVATIONS

The framework presented here deals only with the dyad of mobilized blacks under ANC's leadership and the NP-dominated South African *apartheid* state. The model was applied to analyze the situation that eventually led to the peaceful transfer of power and the disintegration of the *apartheid* state. The model also throws light upon the evolution of the bargaining and negotiations between the two sides in 1990–1994.

This case is of particular interest because of the incidence of violence already occurring by 1989 and the long-standing efforts of the black majority to alter the status quo through both peaceful and violent means. Despite decades of severe discrimination of the majority against the population, institutionalized ideological racism on the part of the state, and on-and-off unrest and attempted insurgency, the two sides (ANC and the *apartheid* state) managed to negotiate a peaceful transition of power.

Use of the model in reference to the last stages of the *apartheid* state illustrates the following main points:

- Long-term trends of increasing international isolation, combined with the transformation of South Africa's economy, increased the vulnerability of the *apartheid* state and led some whites to question *apartheid* on economic grounds;

- Intrawhite political realignment gave rise to new leadership within the *apartheid* state that was no longer tied to the cause of militant Afrikaners and *apartheid;*

- The fall of communism drove both the ANC and the *apartheid* state toward a negotiated solution to *apartheid* by stripping away some of the fears of majority rule among whites and by reducing some external aid and support to the ANC;

- The ANC's skillful leadership, multiethnic platform, and wide appeal overcame the very real possibility of disintegration of the transition process;

- By the late 1980s, the preferred strategy of the *apartheid* state vis-à-vis the ANC changed from a preference for violent repression to a strategy of exploitation that opened the possibility of negotiation;

- The ANC preferred strategy was also one of exploitation, with violence used just enough to make the organization credible, without crossing over the threshold of a mass uprising;

- Potential certainly existed for violence in the dyadic encounter between ANC and the *apartheid* state (hedging strategy for both), but the more interesting observation is that, despite a virtual state of war between them, neither side preferred to resort to violence;

- Amidst the various possible mobilizing group and state types with strong proclivities toward violence identified by the model, the specific group and state types existing in South Africa in 1989 were the least prone to violence.

In retrospect, the strategic choices identified with each side were followed closely as events unfolded in 1990–1994. In that sense, the accuracy of the model was validated with respect to the transition in South Africa. The model captured well the strategic preferences of the two actors. Moreover, although the analysis undertaken here used April 1989 as the cutoff date for data gathering, the same analysis could have been undertaken with a cutoff date in mid-1988, or perhaps even in mid-1987, with similar outcomes. Although 1987 was an early point to consider the initiation of the transition process, many of the crucial factors that drove both sides to negotiations were already in place and should have appeared as important elements of an intelligence assessment of that time. Also, the capacity of the state did not change substantially between 1987 and 1989, and indications of an intrawhite political realignment became evident in the May 1987 elections. Thus, the model might have provided a useful tool for intelligence analysts thinking about the potential for a transition away from the *apartheid* system and toward majority rule in South Africa.

Just as in the case of the Yugoslav chapter, the use of the model to examine the events that led to the South African transition amounts to far more than a retelling of what is by now a well-known story. In addition to serving as a second crucial retrospective test of the methodology of the process model, this case and the analysis contained herein presents another clear illustration of the linkage between specific goals, the resources amassed, and the strategies and choices open to both the mobilized group and the state. The more accommodative strategy pursued by the still NP-dominated state in 1989–1990 stemmed from indications that the previous policy was heading to a dead end as well as the NP's changing base of support, which led to the rise of new leadership that explicitly acknowledged the dead end of the previous policies. In the words of then-President F.W. de Klerk, "If our old policy, which was so unpopular in many circles, could work, then we would have surely clung to it. But as responsible leaders charged with the government of the country, we came to the conclusion that the policy we had planned could simply not work."[24]

The fact that South Africa did not dissolve into strife but managed to retain a democratic political system remains a stunning accomplishment of the negotiators of the transition. As the model indicates, violence-prone solutions easily could have unfolded, depending on any number of factors. But the model also shows how, in a situation with strong potential for severe violence, the specific matchup of group and state types was a most fortunate one, the least prone to violence. This occurred because the resource base of the two sides determined the range of choices between accommodation and strife. Both sides were not averse to strong negotiating tactics and threats because both negotiated from fairly strong positions. For example, despite some economic problems, the *apartheid* state could have relied on outright force and probably could have tried to muddle through for many years. The ANC's declining base of foreign support even may have made such a state strategy relatively successful. On the other hand, the potential problems of resource base notwithstanding, the ANC had unquestioned widespread popularity and influence in South Africa, and the regime's attempt to muddle through would not have changed that fact. But since both sides also

---

[24]Interview with F.W. de Klerk, *BBC Summary of World Broadcasts*, May 9, 1990.

realized their vulnerabilities, the muddling through strategy did not take place.  The ultimate choices stemmed from a rational understanding of the available options.

## ANNEX: DEMOGRAPHIC CHARACTERISTICS OF SOUTH AFRICA IN THE LATE 1980S

The following information about political and population character-istics is based upon the situation in South Africa in the late 1980s. The information presented here is the basic reference for the analysis in this chapter.

**Name:**  The Republic of South Africa (RSA).

**Capital:**  Pretoria.

**Nature of government:**  Oligarchy, with political rights defined by racial categories.  A consociational democracy, dominated by one group, within the oligarchic system; a tricameral parliament pro-vided for political participation for the "whites," "coloureds" (term denoting people of mixed race), and "Asians."  Executive authority was vested in the Office of the President.  The three houses chose the president, with the white chamber of parliament having an outright majority of the votes.

**Organization of the state:**  A republic, with four provinces:  Cape, Natal, Transvaal, and Orange Free State.  In addition, there were ten Bantustans ("historic homelands") set up for the black population: Bophuthatswana, Ciskei, Transkei, Venda, Gazankulu, KaNgwane, KwaNdebele, KwaZulu, Lebowa, and Qwaqwa.  The first four were recognized by the government of South Africa as sovereign states (no other country recognized them as such).  The other six had a status of "self-governing" territories.

**Date of constitution:**  1984.

**Population:**  approximately 35,094,000 (July 1988).  1980 census showed 28,087,489; 1991 census showed 37,944,000.

**Major ethnic groups:** Legal separation and classification of people into four main "racial" groups: "whites" (denoting those of indigenous European background), "blacks" (denoting those of indigenous African background), "Asians" (denoting those of indigenous Asian background), and "coloureds" (denoting those of mixed "race," usually of "black" and "white"). The main division among the whites was into two groups along linguistic and ancestry lines: the Afrikaners (descendants of Dutch, French Huguenot, and German settlers) and the English speakers (primarily descendants of British settlers); a more recent (1970s) group of whites included Portuguese who fled the former Portuguese colonies in Africa. The coloureds were a catch-all category and included the long-established "Cape Coloureds." The category of Asians generally referred to people primarily of Muslim and Hindu background descended from laborers from India and the East Indies, though it also applied to some descendants of Chinese indentured laborers. The black population belonged to a large number of southern African groups: Zulu, Xhosa, Northern Sotho, Southern Sotho, Tswana, Shangana-Tsongo, Transkei, Venda, and numerous smaller groups.

**Languages:** Official languages: Afrikaans and English. The black population usually used their own tribal-based languages. These included Nguni languages (mainly Zulu, Xhosa, Swazi, South Ndebele), Sotho languages (Seshoesho [southern], Sepedi [northern], Tswana [western]), Tsongo, and Venda. The "Cape Coloureds" generally spoke Afrikaans and were often bilingual. Urban "Asians" generally spoke English as well as their own Asian languages. Generally, the language of interethnic communication was English in the urban areas and Afrikaans in rural areas. Among the laborers in the mines, a pidgin language, Fanakilo, was used.

**Religions:** Christianity (in almost all the European Christian denominations), tribal religions, Islam.

**Population statistics:** According to one of the *apartheid* laws, the Population Registration Act of 1950, all South Africans were to be registered at birth according to "race," with four main racial distinctions: white, black, Asian, and coloured. The term "race," as used in *apartheid*-era South Africa, had a meaning akin to rigidly and ascriptively defined ethnicity.

Population statistics are problematic, in that South African census estimates stopped including data for the population of the four "sovereign" homelands (Bantustans) when the South African government no longer considered the territories a part of South Africa. The government also no longer considered the people connected to the "homelands" to be citizens of South Africa, even though large portions of the nominal homeland populations continued to reside and work in South Africa proper. Tables 4.15 and 4.16 show the difference in population figures depending on whether the black populations associated with "sovereign" homelands were counted or not. In any event, the official figures probably undercounted the black population because of widespread reluctance among blacks to provide census data. In reality, the black population was probably larger.

**Table 4.15**

**Population of South Africa (Including "Homelands") by Official "Races"**

| Group | Population (1,000s) | Percent of Total |
|---|---|---|
| Whites | 4,961 | 14.8% |
| Blacks | 24,901 | 74.1% |
| Coloureds | 2,881 | 8.6% |
| Asians | 878 | 2.6% |
| Total | 33,621 | 100.0% |

Percentages may not add up to 100 due to rounding.
SOURCE: *Survey of Race Relations—1986,* Johannesburg:    South African Institute of Race Relations, 1986.

**Table 4.16**

**Population of South Africa (Excluding "Homelands") by Official "Races"**

| Group | Population (1,000s) | Percent of Total |
|---|---|---|
| Whites | 4,574 | 19.5% |
| Blacks | 15,200 | 65.0% |
| Coloureds | 2,810 | 12.0% |
| Asians | 821 | 3.5% |
| Total | 23,391 | 100.0% |

SOURCE:  Compilation of 1985 South African census estimates.

There was substantial ethnic heterogeneity in the intraracial categories, especially among the black population. However, it is difficult to put together a complete picture of the intrablack population breakdown for all of South Africa because population statistics for South Africa did not include the four "sovereign" homelands and the homelands themselves kept poor statistics and/or were not "homelands" to one particular group. Table 4.17 provides an indication of the intrablack distinctions within South Africa proper (excluding the "homelands"). It provides information for the *de jure* populations, meaning those who did not lose South African citizenship, whether they lived in South Africa proper or not. The *de facto* populations were actually much larger, since the same people continued to reside and work in South Africa proper even if they were considered citizens of a "sovereign" homeland. The populations of the six non-"sovereign" but "self-governing" territories are included in the figures for Table 4.17.

Table 4.18 complements these figures by providing information on the population of the four "sovereign" homelands.

Among the whites, intragroup distinctions were based primarily on linguistic affiliation, ancestry, and religion. The Afrikaners considered themselves to have descended from the original Dutch colonists, and almost all were affiliated with the Dutch Reformed churches. The Anglophone population, considered to be descended primarily from British settlers, varied in religious affiliation: Anglican (Episcopal), Methodist, Roman Catholic, Presbyterian, Jewish, and Lutheran.

**Table 4.17**

**Major Intrablack Distinctions Among the Population of South Africa (Excluding *de jure* "homelands" populations)**

| Group | Population (1,000s) |
|---|---|
| Zulu | 5,683 |
| Xhosa | 2,987 |
| Sepedi (Northern Sotho) | 2,348 |
| Seshoeshoe (Southern Sotho) | 1,742 |
| Tswana | 1,357 |
| Shangaan-Tsongo | 996 |

SOURCE: 1980 South African census data.

**Table 4.18**

**Population of the "Sovereign" Homelands**

| "Homeland" | De jure Population (1,000s) | De facto Population (1,000s) | Primary Group |
|---|---|---|---|
| Transkei | 3,000 | 1,500 | Xhosa |
| Bophuthatswana | 3,200 | 1,700 | Tswana |
| Venda | 550 | 385 | Venda |
| Ciskei | 2,200 | 682 | Xhosa |

SOURCE: 1985 individual "homelands" estimates.

**Table 4.19**

**Intrawhite Distinctions Among the Population of South Africa**

| Native Language | Population (1,000s) | Percent of Total Whites |
|---|---|---|
| Afrikaans | 2,581 | 57.0 |
| English | 1,763 | 38.9 |
| Portuguese | 70 | 1.5 |
| others | 114 | 2.5 |
| Total | 4,528 | 100.0 |

Percentages may not add up to 100 because of rounding.
SOURCE: 1980 South African census data.

Most coloureds were members of the Dutch Reformed or Anglican church. Most Asians were either Hindu or Muslim. The most important native languages among the Asians were Tamil, Hindi, Gujarati, Urdu, and Telugu.

# Chapter Five

# THE ETHIOPIAN PROSPECTIVE CASE

## Sandra F. Joireman and Thomas S. Szayna

## INTRODUCTION

This chapter applies the "process" model for anticipating the incidence of ethnic conflict to the potential case of the emergence of ethnically based violence in Ethiopia. Therefore, in contrast to the two case studies presented previously, this case entails the use of the model to consider prospectively the likelihood of an ethnic mobilization turning violent. The chapter examines the case of Ethiopia from the perspective of what an intelligence analyst might conclude were she to use the "process" model. In essence, we look at the potential ethnic mobilization of the Amhara against the Tigray-dominated Ethiopian state structures in an attempt to alter the political arrangements governing Ethiopia more in favor of the Amhara. Data available as of 1997–1998 were used to conduct this analysis.

The choice of Ethiopia as a case study is in no way meant to suggest that this country is somehow predetermined to slide into ethnic tensions or strife. The choice was made on the basis of geographical diversity (as other case studies examine different regions of the world) and the country's regional importance.

Through the use of the model, the chapter examines the potential grievances the Amhara might have against the state and the likely path that their mobilization might take. Two secondary scenarios examine the potential ethnic mobilization of the Oromo and the Somali against the Tigray-dominated state, the potential paths of the two mobilization processes, and the likelihood of strife that may result.  This orientation is in line with the focus of the "process" model on one particular kind of ethnic conflict, namely, the rise of an ethnic group challenging the state.

The reason for selecting the Amhara challenge as the main scenario stems partly from the specific influence that the Amhara have had in Ethiopia in the past, their substantial population, and some evidence of ongoing attempts at mobilization.  The secondary scenarios examine the other main ethnic group (the Oromo) and a group (the Somali) that is minor in terms of population but historically troublesome to Ethiopian state authorities.  The authors emphasize that, at this point of its development, the model is intended to be more suggestive than predictive.  Neither the choice of the scenarios nor the potential outcomes examined should be taken to imply that the groups in question will mobilize against the government or that confrontation with the state necessarily will result in violence.

As presented in the model, the group-versus-state conflict is the simplest form of ethnic competition and, some may argue, not realistic when applied to the multiethnic conditions of Ethiopia.  The authors do not deny the obvious influence that any open Amhara-Tigray competition may have on the behavior of the Oromo toward the state and other ethnic groups in Ethiopia.  However, the authors believe that, for the purposes of outlining in a parsimonious manner the basic preferences that the state and the Amhara would have, the model is sufficient.  And the model is sufficiently flexible to allow taking into account other groups' impact upon the dyad in conflict if the impact is likely to be major.  Finally, it is not clear whether the advantages of a multigroup model would outweigh the complexity and probable difficulty of its use.

Following the organization of the previous two chapters, four sections follow this introduction.  The first section examines the structure of closure and provides an analysis of which ethnic groups appear to be privileged and which seem underrepresented in structures of power (the demographic characteristics of Ethiopia in 1997–1998, on which the analysis is based, are appended at the end of the chapter).  Then it examines the strengths and weaknesses of the challenging ethnic group—the Amhara—by looking at its potential mobilization process.  A shorter analysis of the secondary scenarios is included.  The second section looks at the capabilities the state—

federal Ethiopia—might bring to bear in dealing with the challenging groups. The third section examines the strategic choices, arrived at on the basis of the assessments in earlier sections, that the state and the group are likely to pursue vis-à-vis each other, given their resource bases. Finally, a few observations conclude the chapter.

## ASSESSING THE POTENTIAL FOR STRIFE

### Closure in the Political and Security Realms

Since 1991, the Ethiopian People's Revolutionary Democratic Front (EPRDF)—the current party of government in the Ethiopian Parliament—has made an effort to open cabinet-level leadership and the composition of the army to other ethnic groups, but such measures are widely perceived as less than genuine or an outright sham. When the EPRDF won the Ethiopian civil war in 1991, it was a predominantly Tigrayan organization (having recently changed its name from the Tigrayan People's Liberation Front, or TPLF).[1] One method the EPRDF used to broaden its ethnic representation was to establish "friendly" coalition parties, drawing representation from each ethnic group. However, the ethnic parties are not necessarily representative of their respective ethnic groups. For example, the Oromo People's Democratic Organization (OPDO), formed to represent the Oromo people within the EPRDF coalition, was originally composed of captured Oromo POWs and officers. The OPDO is viewed with suspicion by many Oromo because of its origins and closeness to the EPRDF. Its main rival in representing the interests of the Oromo people is the Oromo Liberation Front (OLF), a much older and well-established resistance army and political party.[2] This pattern of newly established EPRDF-affiliated groups, challenging older, rival parties to represent ethnic interests, is repeated in other areas.

Regional and national elections of Ethiopia's highest-ranking political authorities took place in May and June 1995. The president has

---

[1] For a history of the TPLF, see John Young, "The Tigray and Eritrean Peoples Liberation Fronts: A History of Tensions and Pragmatism," *Journal of Modern African Studies,* Vol. 34, No. 1, 1996, pp. 105–120.

[2] There are also other smaller groups such as the Oromo National Liberation Front, which draw support from the Oromo population.

mainly ceremonial powers and is elected by the Council of People's Representatives. The prime minister holds most of the executive functions. The prime minister, elected by the Council of People's Representatives, appoints the 17-member Council of Ministers (subject to approval by the legislature). See Tables 5.1 and 5.2.

Representation within the highest levels of the government—the Council of Ministers and the legislature—is ethnically diverse, but the true nature of power sharing is not clear. See Table 5.3. At least five Tigrayan assistant ministers may have direct access to the prime minister, which casts some suspicion on the true integration of the

### Table 5.1

### Top Ethiopian Officials

| Position | Name | Ethnicity |
|---|---|---|
| Prime Minister | Meles Zenawi | Tigrayan |
| President | Negasso Gidada | Oromo |
| Secretary of the Council of the People's Representatives | Hailu Hailifom | Tigrayan |
| Speaker of the Council of the People's Representatives | Dawit Yohannes | Amhara |

### Table 5.2

### Council of Ministers, Ethnic Breakdown

| Ethnic Group | Number of Ministers | Percent of Ministers |
|---|---|---|
| Oromo | 4 | 24 |
| Amhara | 4 | 24 |
| Gurage | 2 | 12 |
| Tigray | 1 | 6 |
| Harari | 1 | 6 |
| Kembata | 1 | 6 |
| Somali | 1 | 6 |
| Afar | 1 | 6 |
| Weleyta | 1 | 6 |
| Hadiya | 1 | 6 |

council.[3] In the legislature, the representation of various ethnic groups is large, but the overwhelming majority of the members of parliament (slightly over 90 percent) are from EPRDF-affiliated parties. Each region, which is ethnically defined, is allocated a certain number of seats within the parliament. Typically, political parties nominally controlled by local ethnic groups but under the EPRDF umbrella have taken these seats. This is due, at least in part, to the opposition boycott of the 1995 parliamentary election, in which most of the opposition parties declared themselves unwilling to participate.[4] Thus, the parliamentary seats have been filled with people representing their particular ethnic group, yet there is substantial room for doubt as to just how much these MPs are trusted by most members of their ethnic groups.

### Table 5.3

### Representation in Government by Linguistic Affiliation

| Language | Speakers (100,000s); Percent of Total | Number and Percent of House Seats Reserved for Their Region (n = 548) | Representation by Political Party; Percent of Total House |
|---|---|---|---|
| Amhara | 15 (28.5) | 147 (27) | 144 (26) |
| Oromo | 14 (26.6) | 187 (34) | 182 (33) |
| Tigray | 4 (7.6) | 40 (7.3) | 40 (7.3) |
| Weleyta | 2 (3.8) | NA | 13 (0.4) |
| Somali | 2 (3.8) | 25 (4.6) | 0 |
| Gurage | 1.5 (2.9) | NA | 19 (3.5) |
| Hadiya | 1 (1.9) | NA | 9 (1.6) |
| Kembata | 1 (1.9) | NA | 5 (1.0) |
| Afar | 0.5 (1) | 8 (1.5) | 8 (1.5) |
| Harari (Aderi) | 0.03 (0.06) | 1 (0.2) | 1 (0.2) |

---

[3]"Looking Federal," *Africa Confidential,* September 22, 1995, p. 5.

[4]Their motive for doing so seems to have been threefold: (1) to withhold legitimacy from the current government, (2) to protest restrictions placed on opposition access to the print and television media, and (3) to protest the repression that some opposition parties faced in the more remote areas of the country. See Terrence Lyons, "Closing the Transition: The May 1995 Elections in Ethiopia," *Journal of Modern African Studies,* Vol. 34, No. 1, 1996, pp. 121–142. Lyons documents reports of restrictions on OLF political activities in the countryside.

Data on the ethnicity of national director-level managers are not available. Before the regime change Amhara were overrepresented in such positions, and that situation may still persist. But based on the policies of the government to equalize opportunities among ethnic groups, there may be a lessening of the Amhara influence over the director-level management of national institutional structures.

The opposite is true at the regional level, where information is available. In the regions there is a significant bias toward employing people of each region's dominant ethnic group. For example, in the Southern People's Region there has been an attempt to promote members of local ethnic groups into the offices of judge and prosecutor.[5]

In terms of closure in the security realm, top positions are dominated by Tigrayans and Amharas. The top security authorities in the country are the prime minister, Meles Zenawi, who both controls his own personal security retinue and serves as the commander-in-chief of the armed forces; the acting minister of justice, Worede-Wold Wolde,[6] who controls the police force; and the minister of defense, Teferra Walwa, an Amhara who supervises the armed forces. Kinfe Gebremedhin, a Tigrayan and the former chief of security, controls the Security, Immigration, and Refugee Affairs Authority, which previously was the Ministry of Internal Security. Kinfe Gebremedhin reports directly to the prime minister, Meles Zenawi.[7]

The minister of defense, Teferra Walwa, the highest-ranking security official, is a member of the Amhara National Democratic Movement (ANDM, an Amhara party under the EPRDF umbrella). He replaced the previous minister of defense Tamirat Layne, who was fired very publicly on charges of corruption. Beyond the ministerial level, there is little information available about key security officials. In addition, the administrative responsibilities of some of the security agencies are not clear. For example, the role of Kinfe Gebremedhin at the benignly labeled Security, Immigration, and Refugee Affairs Authority remains to be fully understood and yet is a key to determining the

---

[5]S. F. Joireman, interview with SPR court officials, Yirgalem, 1994.

[6]Justice Minister Mahetema Solomon resigned in June 1997.

[7]"Looking Federal," 1995.

relative power and control of the security apparati.  As the former chief of security for the TPLF and as someone with direct access to the prime minister, his role in internal security affairs most likely equals or surpasses those of the ministers of defense and justice. Indeed, his placement as head of the agency responsible for internal security is an indication that the prime minister has viewed the greatest threat to the Ethiopian state to be internal rather than external.

Similarly, the ethnic makeup of the individuals who form the upper levels of the security apparati is not publicly known.  The Ethiopian military has made well-publicized efforts to diversify the ethnic composition of the lower levels of the armed forces, but it is not clear to what extent, if any, this diversification has occurred in the officer corps.  In 1995, only three of the military's nine generals were Tigrayan.[8]  However, Tigrayans are known to be widely represented in the officer corps.

**Assessment of closure in the political and security realms.**  Closure in Ethiopia is based more on political opinion than ethnicity.  The government has attempted to incorporate members of different ethnic groups into its political fold, but only insofar as they agree with the government platform and agenda.  The Oromo provide an important illustration.  Those Oromo affiliated with the pro-government faction, the OPDO, are represented in national and regional governments.  Those in the anti-government faction (OLF members and supporters) are not represented in either chamber of the legislature and did not stand for election.  We have no way of measuring support for the two Oromo factions.  However, there are reports that at least some Oromo view the OPDO as "seriously compromised" due to its affiliation with the EPRDF.[9]

Unfortunately, information on the exact ethnic composition of the Ethiopian armed forces, both enlisted and officers, is not readily available.  Additionally, there are no data on the composition of the regional police forces.

---

[8]U.S. Department of State, "Ethiopia Country Report on Human Rights Practices for 1996," Released by the Bureau of Democracy, Human Rights and Labor, January 30, 1997, *http://www.state.gov/www/issues/human_rights/1996_hrp_report/ethiopia.html.*

[9]Lyons, "Closing the Transition," 1996.

Since there are no formal restrictions on access to power in political or security spheres, the static snapshot of the Ethiopian system presented above appears to be dynamic enough to accommodate change. Indeed, the government has gone out of its way to be inclusive of individuals of other ethnic groups willing to cooperate with its political agenda. Ethnically based parties opposed to the EPRDF, however, face informal government pressure, general harassment, and a denial of access to the media.

A formal process for the replacement of political elites—including a challenge along ethnic lines—exists, but opposition groups have not taken the opportunity, restricted though it may be, to express their dissent through the electoral process. Opposition groups have repeatedly refused to participate in elections, the most recent of which occurred in 1995.[10] This has placed both the government and the country in an odd position. There is a means to replace the current ruling group, but thus far the opposition has made no concerted attempt to take advantage of it; since the means has not been tested, its genuineness is unclear.

There are informal barriers to political participation. There is some suppression of overt dissent in Ethiopia.[11] The threat of government intolerance to political opposition is real, highlighted by the assassination or imprisonment of vocal opposition leaders within Ethiopia. The government appears willing to tolerate quiet and reasoned dissent, but when the opposition becomes passionate, the authorities crack down.

There are no formal rules restricting the access of any ethnic group to power at a national level in Ethiopia. But informally, EPRDF has made no secret of its desire to keep a monopoly on power. Moreover, as stated earlier, within the regions there is an explicit preference for officials from ethnic groups indigenous to the regions.

---

[10]S. F. Joireman, "Opposition Politics in Ethiopia: We Will All Go Down Together," *Journal of Modern African Studies,* Vol. 35, No. 3, September 1997.

[11]There have been two notable instances of repression of dissent under the current government, both involving demonstrations by university students. The first occurred in 1991 shortly after the EPRDF entered the capital, and the second occurred in March 1997. "University Students Released," *Addis Tribune* (Addis Ababa), March 28, 1997.

Individuals from minority groups within regions will experience barriers to government or administrative services.

## Closure in the Economic Realm

There is very little subnational data available on the economic sector in Ethiopia. The results of the 1994 census, if they are ever fully published, may shed more light on the topic. Agriculture is the foundation of the Ethiopian economy. It accounts for approximately 43 percent of the gross domestic product and most of the country's export income. It is the primary source of employment in the country: 80 percent of Ethiopians are engaged in it. Apart from farming, the state apparatus and the military are the next-greatest sources of employment in the country.[12]

There is no explicit information available on the distribution of income within Ethiopia. Keeping in mind that neither farming, working for the government, nor being a soldier pay particularly well, we can guess that the upper tier of the Ethiopian economy is composed of people who are engaged in business pursuits. We can also posit that the wealthiest in the society are urban Amhara (because the majority of the educated, urban dwellers in Ethiopia are Amhara).[13]

Although there are no official figures available on the distribution of wealth in the population as a whole or with regard to ethnicity, the population of Ethiopia is generally extremely poor. The World Bank's estimate of the Ethiopian GNP per capita in 1996 (latest available) was $110, placing the country among the poorest in the world, at about the level of Somalia and Bangladesh.

However, we can make a few general observations about relative wealth. First, because coffee is grown as a cash crop in the south,

[12]World Bank, *From Plan to Market: World Development Report 1996*, New York: Oxford University Press, 1996, p. 188.

[13]Recent census information puts the Amhara population at 49 percent of the total population of Addis Ababa. Transitional Government of Ethiopia Office of the Population and Housing Census Commission, Central Statistical Authority, *The 1994 Population and Housing Census of Ethiopia Results for Addis Ababa*, Volume I—Statistical Abstract, Addis Ababa: Central Statistical Authority, 1995.

southern peasants who grow coffee are slightly better off than their northern counterparts. Those lucky enough to have a college education or a job in the capital city or another urban center are better off than the farmers in the countryside. With improvements in administration and efficient reinvestment of resources, the Southern Nations and Oromo Regions should be the most wealthy in the country due to the coffee and tea production there.

**Assessment of closure in the economic realm.** Again, it is extremely difficult to make any conclusive statements on the structure of wealth in Ethiopia. The only statement that is supportable (and only on the basis of anecdotal evidence) is that because of their overrepresentation in the urban areas and because of the preferential access they have had to education over the past century, the elite of Ethiopia tend to be Amhara.

Is there a mechanism at hand for changing the static snapshot presented above? There are no formal restrictions on access to wealth in Ethiopia. The only barrier to economic activity that is enshrined in the political structure (albeit informally) is the restriction on agricultural investment in regions other than the home region of a particular ethnic group.[14] Additional problems may develop in the realm of land allocation and land reform. There is currently no market for land in Ethiopia, and regions are in the process of developing policies for land allocation and transfer. Should these policies tend to exclude minority groups in the regions, there would be tremendous potential for conflict. There are no identifiable restrictions on wealth access that may disadvantage individuals of a particular ethnic group.

## Closure in the Social Realm

Status distinction in Ethiopia is a prominent issue, and both status and ethnicity have become more controversial since the 1991 regime

---

[14]Sarah Gavian and Gemechu Degefe, "Commercial Investors and Access to Land," in Dessalegn Rahmato (ed.), *Land Tenure and Land Policy in Ethiopia after the Derg*, Trondheim, Norway: Working Papers on Ethiopian Development, No. 8, 1994.

change.[15] Status at the national level only partly reflects political dominance. The EPRDF is predominantly Tigrayan, but the Amhara still retain the preeminent social position. A status stratification map for Ethiopia might work out as illustrated in Table 5.4.

The Amhara and Tigray fill the top two places of the stratification map for three historical reasons: nobility, mobility, and Christianity. Both groups trace their ancestry back to the Abyssinian Empire of the 12th century AD.[16] Their joint cultural myth extends back even further, dating the origin of these peoples to the meeting between King Solomon and the Queen of Sheba (an Abyssinian).

Mobility also explains the dominance of the Amhara and Tigray peoples. In the 19th and 20th centuries the two groups moved out of their traditional lands in the northern highlands of Ethiopia and took over a large portion of the fertile coffee lands in the south. This

### Table 5.4

### Status Stratification

| Ethnic Group | Status |
| --- | --- |
| Amhara | + + |
| Tigrayan | + + |
| Aderi/Harari | + |
| Gurage | + |
| Oromo | – |
| Somali | – – |
| Afar | – – |

+ + = high status; – – = low status (compiled on the basis of data presented earlier).

---

[15]The reasons are primarily tied to the "nation-building" endeavor by the government. John Markakis, "Ethnic Conflict and the State in the Horn of Africa," in Katsuyoshi Fukui and John Markakis (eds.), *Ethnicity and Conflict in the Horn of Africa*, Athens, OH: Ohio University Press, 1994, pp. 217–237.

[16]For further information on Ethiopian political history, see Edmond J. Keller, *Revolutionary Ethiopia: From Empire to People's Republic*, Bloomington: Indiana University Press, 1988; and Harold G. Marcus, *A History of Ethiopia*, Berkeley: University of California Press, 1994.

movement into the southern marches of the Ethiopian Empire resulted in the enserfment of the peoples who already lived on the land. Those enserfed were the Oromo, Gurage, Gedeo, and Wolaita people, among others. When the Amhara and Tigrayans moved into these areas they brought the apparatus of the state with them and enforced tax collection and the speaking of Amharic as the language of administration within the country.

The third reason that both the Amhara and Tigray peoples are at the higher end of the stratification map is that both groups are predominantly Orthodox Christian. This gives them a higher status due to the ancient history of Orthodox Christianity in the region. The presence of the religion has meant that there has been a written record in Ethiopia since the 12th century, along with manuscripts on both religion and politics. This ancient record is combined with a modern sense of the importance of a "Christian" Ethiopia as a bulwark against Islamic influence.

Until recently, the Amhara had unquestioned social dominance. There was a separation between the Tigrayans and the Amhara during the early part of the 20th century as fighting developed between Ras Hailu, the Tigrayan aspirant to the throne, and Haile Selassie, an Amhara. As Haile Selassie's rule solidified, so did the iron grip of the Amhara over the political and economic institutions of the country. When the revolution came in 1974, Haile Selassie was succeeded by Mengistu Haile Mariam, who, though of mixed parentage, identified himself with the Amhara. Mengistu ruled until 1991.

The long stretch of Amhara leaders in Ethiopia has led to two inaccurate and insidious ideas among the Amhara. The first is the association of a particular Amhara "type" with the definition of Ethiopian: by this notion, an "Ethiopian" is a slight, light-skinned, Christian, Amharic-speaking farmer. The second is a sense of manifest destiny: an idea that the Amhara are somehow ordained to rule a "Greater Ethiopia."[17] These ideas are insidious because they give no quarter to the majority of the population who do not fit this "type" linguistically, religiously, or ethnically.

---

[17]"Greater Ethiopia" is a term used to include Eritrea, the state to the north of the current Ethiopian border (along the Red Sea Coast), that recently gained its independence.

It used to be the case that Amharans saw the Tigrayans in a consanguineal fashion, but political differences made ethnic distinctions grow. Tigrayans spoke a different language, but it was understandable to Amharic speakers, and their cultures were virtually indistinguishable. The rift between the two groups was exacerbated in the mid-1980s, when the Tigrayan People's Liberation Front (TPLF) became a thorn in the side of the Mengistu regime. As the TPLF began its advance on the capital city of Ethiopia and then took it over in 1991, the Tigrayans ceased to be cousins and became "foreigners" in the rhetoric of the radical Amhara.

The Aderi, also known as the Harari, are aligned with the Tigrayans on the stratification map, just as they are in government. The Aderi are a population of approximately 26,000 who reside in and around the ancient Islamic city of Harar in southeastern Ethiopia. They have a high social status within the country because they are regarded as distinct, refined, and ancient in their origins. The Aderi are overrepresented in parliament and the Council of Ministers, as they are given one representative in each body for a group that is 0.0005 percent of the Ethiopian population. Other groups, with larger populations, don't have any representatives in the Council of Ministers. This overrepresentation is due to the particular status of Harar, along with Addis Ababa, as a designated multiethnic city-state, independent of any regional government.

The Oromo and Gurage are both historically oppressed and predominantly Islamic southern peoples.[18]  Amhara and Tigrayan northerners enserfed large numbers of Oromo peasants in southern Ethiopia until 1974. Even after 1974, the Oromo and Gurage were forced to learn Amharic in order to achieve any sort of educational or vocational success. The Gurage are slightly higher on the social

---

[18]There are an increasing number of studies on the history of the Oromo and Oromo nationalism. Asafa Jalata, "Sociocultural Origins of the Oromo National Movement in Ethiopia," *Journal of Political and Military Sociology,* Vol. 21, 1993, pp. 267–286; Asafa Jalata, *Oromia and Ethiopia: State Formation and Ethnonational Conflict, 1868–1992,* Boulder, CO: Lynne Rienner, 1993; P.T.W. Baxter, "The Creation and Constitution of Oromo Nationality," in Fukui and Markakis, *Ethnicity and Conflict in the Horn of Africa,* pp. 167–186; Edmond J. Keller, "The Ethnogenesis of the Oromo and its Implications for Politics in Ethiopia," *Journal of Modern African Studies,* Vol. 33, No. 4, December 1995, pp. 621–634; and Asafa Jalata, *Oromo Nationalism and the Ethiopian Discourse: The Search for Freedom and Democracy,* Trenton, NJ: Red Sea Press, 1997.

stratification map than the Oromo because they are associated with positive qualities based on their perceived business acumen.

The Somali and Afar peoples bring up the lower segments of the stratification map because, in addition to being predominantly Islamic, they are pastoralists. Pastoralism is viewed in Ethiopia and throughout Africa as being less prestigious (even less human) than settled agricultural practices.

What are the implications of this status distribution? At the national level, there are no tangible benefits or disadvantages to being of a particular ethnicity. In the regions there is a specific preference for those of the local ethnicity to fill positions in the regional administration. Linguistic groups are not discriminated against in their regions, but non-Amharic speakers may find themselves disadvantaged if they interact with the bureaucracy of the national government. Informally, a whole range of biases and perceptions surrounds status hierarchy, with members of the less urbanized and non-Christian groups seen as less sophisticated or cultured. Such biases have a self-perpetuating and self-reinforcing component and affect myriad everyday decisions.

The fact that there is very little movement between status groups attests to the lack of a mechanism for change within the Ethiopian social stratification system. Indeed, the norm for social tolerance is changing toward a greater degree of intolerance and ethnic solidarity. Since the ratification of the new Ethiopian constitution, ethnic group and ethnic identity have become fundamentally important on a political and social level. Ethnicity defines regional appointments, political sympathies, and party affiliation.

## Overall Assessment of Closure

Based on the information presented above, Table 5.5 summarizes the degree of closure (in an overall sense as well as in the political, economic, and social realms) experienced by Ethiopia's main ethnic groups. To reiterate, closure in Weberian terms refers to the "process of subordination whereby one group monopolizes advantages by closing off opportunities to another group." In the table, a group experiencing a "low" degree of closure has the most opportunities open

**Table 5.5**

**Patterns of Closure by Ethnicity in Ethiopia**

|              | Political | Economic | Social   | Overall       |
|--------------|-----------|----------|----------|---------------|
| Amhara       | Low       | Low      | Low      | Low           |
| Tigrayan     | Low       | Low      | Low      | Low           |
| Aderi/Harari | Low       | Moderate | Low      | Low           |
| Gurage       | Moderate  | Low      | Moderate | Moderate      |
| Oromo        | Moderate  | Moderate | Moderate | Moderate      |
| Somali       | Moderate  | Moderate | High     | Moderate/High |
| Afar         | Moderate  | Moderate | High     | Moderate/High |

to it.  A group experiencing a "high" degree of closure has opportunities largely closed off.

The Amhara and Tigrayans show a consistent pattern of privilege, and the Oromo and (especially) the Somali and Afar peoples are experiencing a moderate degree of closure in Ethiopian social and economic life.  Politically, Tigrayans are dominant, but they do share power to a moderate degree with other ethnic groups.

Nevertheless, poverty is a great equalizing mechanism among the various ethnic groups.  People in the countryside do not suffer high degrees of relative deprivation.  In the urban areas, however, the Amhara and Tigrayans are present in greater numbers, and they are more educated and better employed than the Somali and Afar.  Yet even if we take the differences between ethnic groups in the urban areas into account, the relative deprivation between groups is quite low.

Issues of closure in Ethiopia are primarily historic in their origin, but they are changing, and we should expect to see changes in the economic and social positions of the ethnic groups in this generation.  Indeed, as the EPRDF political institutions take root, we should see some groups, particularly the Oromo, gain in political power. As they become politically powerful, we can expect their social position to improve. Thus, the stratification patterns have some flexibility.

A final ranking of groups along the lines of privileged to dominated—in relative terms—is given in Table 5.6.  The specific placement of ethnic groups on the privileged-dominated scale is not evenly

**Table 5.6**

**Ranking of Ethnic Groups in Ethiopia**

| Privileged | Amhara |
|---|---|
| | Tigrayan |
| | Aderi/Harari |
| | Gurage |
| | Oromo |
| ↓ | Somali |
| Dominated | Afar |

spaced, and the overall differences between the top and bottom rankings are not extreme.

## TRANSFORMING POTENTIAL STRIFE INTO LIKELY STRIFE

As a prospective case study, this section focuses on the process of potential mobilization of one of the main ethnic groups against the state. Three distinct potential processes of mobilization are sketched out, each based on one of the ethnic groups. The primary scenario focuses on the Amhara, and that mobilization process is sketched out in full. The secondary scenarios, focusing on the Oromo and Somali, are discussed in a briefer format.

### PRIMARY SCENARIO: AMHARA MOBILIZATION

The Amhara are among the most privileged ethnic groups in Ethiopia. However, there is a critical disjuncture between Amhara social and economic power and their current political weakness. Resentful of their fall from power, suspicious of the current regime, distrustful of any attempts at inclusion, and possessing a mythology of a "right to rule Ethiopia," the Amhara are in a position to be mobilized for political action along ethnic lines. The abrupt change in power relations in Ethiopia left the Amhara embittered, hostile, and in search of a scapegoat.[19]

---

[19]Scapegoats have been found alternatively in Eritrea and the United States.

## Incipient Changes

The ethnically inclusive declaratory policy of the current government, as well as a strong Tigrayan role in the government, points to longer-term trends under which the Amhara cold see their relative share of power draining away.  These trends put in place an ever-present source of concern among the Amhara for their social and political position in the country.

Sudden or at least less-expected changes, caused by centrifugal tendencies in other regions of Ethiopia, also could spark greater Amhara mobilization.  For example, there have been rumors of a referendum on independence for Oromia.[20]  Although such a referendum would be unlikely to succeed, given the composition of the regional assembly, public discussion of such a thing will almost certainly incite Amhara anger and provide fuel for mobilization.  While many Oromo see the 19th-century Amhara/Tigrayan expansion as colonial, Amhara tend to view "the conquered lands" as an inalienable part of Ethiopia and would resist their dispossession.

## Tipping Events

Galvanizing or tipping events might involve the release of Amhara firebrands from prison or restrictions on Amhara media.  The most famous of the major recent attempts to mobilize the Amhara are Professor Asrat Woldeyes's efforts to incite the rural Amhara to the same level of fervor as radical urban Amhara.  His efforts in this regard led to a jail sentence of five and a half years for attempting to incite ethnic violence.  Professor Woldeyes, viewed by many as the spokesperson for the radical Amhara, faces additional charges associated with a prison escape in 1994, which resulted in the death of several prison guards.  Extended detention or lack of a trial for Woldeyes might prove a galvanizing factor in group mobilization.

A second possible tipping event would be a restriction on the publication of opposition newspapers such as *The Addis Tribune* and other Amharic or Amhara-targeted papers.  Amhara opposition groups are already extremely sensitive to government censorship,

---

[20]"TPLF talks with OLF," *Ethiopian Review*, May–June 1997, p. 12.

due to the fact that many of the journalists who have fallen afoul of the government are Amhara.[21]

## Leadership

Because the urban Amhara tend to be better educated than members of other ethnic groups, there are many present and potential leaders. Some Amhara activist leaders, such as Asrat Woldeyes and Taye Woldesemiat, are already in jail. Both men were university professors dismissed with 39 others in 1991 for their political activities. In addition to these 41 activists are other urban educated Amhara who have been working for the government or in the private sector and have the knowledge and awareness of organizational techniques to be more than able to lead an opposition group. Thus the leadership potential among the Amhara is deep, though the emergence of a specific individual as the leader is difficult to predict.

## Resources and Organization

There seem to be ample resources available to the Amhara relative to other groups (that is, within the overall constraints of an extremely poor country). Radical Amhara groups within Ethiopia draw their support primarily from the population of Addis Ababa. The urban Amhara are among the wealthiest groups in Ethiopia, and they potentially provide a large resource base. There are signs that Amhara activists are already using resources to obtain arms; the Ethiopian government has uncovered one Amhara attempt to cache arms.

Political parties such as the All-Amhara People's Organization provide organizational structures for Amhara dissent and potential political violence. An antigovernment urban terrorist group active in Addis Ababa called the Ethiopian Patriotic Front offers an even more radical vehicle for violence, and it apparently has mastered the

---

[21]The government has imprisoned 15 journalists in 1995–1996 for violation of the press law. Some have been imprisoned for inciting ethnic hatred, but others are in prison for violating accepted standards of decency. "Journalists Under Fire: 27 Dead, a Record 183 Jailed in '96," *The Washington Post,* March 15, 1997.

organizational skills required of a conspiratorial group.[22]  With some of the most educated elite of Ethiopia, the Amhara would have ample access to organizational skills, and would probably be the source of a relatively large share of the overall availability of such skills in Ethiopia.

A chauvinistic Amhara movement has moderate to high organizational potential.  Perhaps its greatest obstacle is that the numerically large Amhara population in the countryside is difficult to mobilize. In addition to the normal conservatism that might be expected from any peasant group, the Amhara peasants also experienced far more of the last civil war than their urban counterparts did and may still perceive the Tigrayan-dominated government as "Ethiopian" rather than ethnic antagonistics.

### Foreign Element

The most significant foreign influence on ethnic relations in Ethiopia comes from Amhara abroad.  The revenues collected overseas for the Amhara cause, particularly in the United States and the Scandinavian countries, are probably the most important source of support for the radical Amhara.  In addition, Amhara groups in the United States give aid to the mobilization process in the form of intellectual justifications against the present status quo, disseminating them in Ethiopia through both hard-copy publications and the Internet. There is currently an effort to raise money in the United States for the Ethiopian Unity Front (EUF), an Amhara insurgency movement.[23]  It is unclear at the moment whether the EUF has started activity within Ethiopia, but the organization appears to have a ready cadre in the United States.

A full-blown Amhara mobilization could count on little support from neighboring governments.  Eritrea does not want to see the Amhara back in power, since it will then become a target of Amhara irreden-

---

[22]Security forces within the capital city have stopped the open activities of this group. The police continue to crack down on its remnants, one of whom, Assefa Maru, was killed while allegedly resisting arrest in May 1997.  "ETA Executive Committee Member Shot," *Addis Tribune* (Addis Ababa), May 9, 1997.

[23]This group has been raising money in various U.S. cities.  It is associated with the Amhara-dominated party COEDF.

tist aspirations (an irony, given the denunciation of Tigrayans as foreigners by radical Amhara). Somalia is no longer a functioning state and does not seem likely to become one in the near term, ruling out its likelihood of being able to spend resources to support a major social movement in Ethiopia. Sudan is something of a wild card. Its Islamic government is unlikely to support any faction in Ethiopia except the Oromo. Sudan currently finds Ethiopia a convenient scapegoat and blames it for the recent offensive of the Sudan People's Liberation Army—Sudan's own ethnic insurgency movement.

## Overall Assessment of Mobilization

Strength of leadership will determine the intensity and effectiveness of Amhara mobilization. Small cell groups that try to spearhead a mobilization have formed already among the Amhara; at least some of these cells are prepared to use violence in the pursuit of their goals. With stronger leaders or, more specifically, leaders able to gain the support of the countryside, the potential for effective and widespread mobilization would escalate greatly. But the other elements of mobilization (resentment over perceived status usurpation, access to resources both at home and abroad, strong organizational potential) are in place, and the Amhara could command a powerful movement.

### SECONDARY SCENARIO: OROMO MOBILIZATION

Numerically the largest ethnic group in Ethiopia by many estimates, the Oromo have never been influential in Ethiopian government and they have a low status in contemporary Ethiopia. The Oromo face a moderate level of closure and can point to some real grievances. The primary problem that sustains the continuing low status and influence of the Oromo is their weak leadership. With a cohesive leadership, the Oromo probably would be the single most powerful ethnic group in Ethiopia. But so far, the leaders of the various Oromo factions have been unable to agree on either goals or tactics.

The Imperial Ethiopian government (before 1974) imposed brutal sanctions on the Oromo as a group. The Oromo obtained access to greater economic, if not political, benefits under the Derg government (1974–1991). In the early 1990s, the Oromo Liberation Front

(OLF) worked with the EPRDF to facilitate the overthrow of the Derg regime. It was there, however, that their trouble with the current government began, as two opposing factions developed within the Oromo. The first faction wanted some type of cooperation and participation with the EPRDF. The second favored secession from the Ethiopian state and the establishment of an independent "Oromia."

The EPRDF's creation of the Oromo People's Democratic Organization (OPDO) under the EPRDF umbrella exacerbated the rift within the Oromo. This group, in collaboration with the EPRDF, drew power away from the OLF. The OPDO, representing an Oromo minority, participated in the elections of 1992 and 1995 and won all of the Oromo seats because the OLF declined to participate. The OLF remains organized in the countryside and retains control of weapons and men. There have been reports in 1996–1997 of OLF-initiated hostilities in the Oromo region. In other words, there is a dedicated cadre—willing to use violence—in place, and it could spearhead the mobilization process. If the rumored referendum on the independence of Oromia were to occur, a more powerful and united Oromo movement would emerge no matter what the outcome of the vote might be.

The splintering and weakening of the Oromo as an ethnic group organized for political ends could all be changed with the appearance of a young and charismatic leader. Youth is a necessity because of the leadership conflicts and struggles for power that have divided the older Oromo leadership. An Oromo mobilization might be aimed at either a substantial share of political power within Ethiopia or an outright independent Oromia.

In an overall sense, the mobilization potential of the Oromo would face resource problems. However, in the extremely poor conditions of contemporary Ethiopia, the pure force of numbers could offset some of the resource problems.

## TERTIARY SCENARIO: SOMALI MOBILIZATION

The Somali population (part of the Ogadeni clan) is numerically small but inhabits a compact area. The Somali face a moderate level of closure and have real grievances against the government. A Somali mobilization would have limited resource potential, especially vis-à-vis the relative resource levels available to the Ethiopian state. But if there were sudden changes in the balance of power in Ethiopia (arising, for example, from one of the two scenarios described above), the mobilization might be successful. Foreign assistance—from Somalia—is a crucial variable that could augment substantially the resource base of the Somali in Ethiopia.

Since the British relinquished the territory in 1955, the Ogaden has been the focus of several military conflicts between Ethiopia and Somalia (1960, 1964, 1973, 1977, and 1996).[24] Since 1991, the Ethiopian government has engaged the armed Ethiopian Somali group Al-Ithad Al-Islam in the Ogaden and across the border in Somalia. The 1992 elections in the area had to be postponed because of regional insecurity. The goal of the Al-Ithad Al-Islam—a group that could lead the mobilization process—is openly secessionist, as the group has tried to detach the Ogaden, now called the Somali region, from the Ethiopian state. The Al-Ithad group has been implicated by the government of Ethiopia in several bombings in 1996–1997 in the capital city of Addis Ababa.[25]

Currently, the Ethiopian government maintains control of the Ogaden through administrative and police means, keeping any mass mobilization at bay. But if the Somali state were to again become viable, the situation might change. Moreover, if the Somali state were to resurface with a ruler from the Ogadeni clan, the stability of the Ethiopian Somali region certainly would be at risk. The area also may have potential for oil production. It was, in fact, rumors of oil discoveries in the Ogaden that led to the 1973 conflict between Ethiopia and Somalia. Further discoveries or rumors of discoveries may exacerbate the conflict that already exists.

In an overall sense, the mobilization potential of the Somali is constrained by resource problems. But the situation could change in the relative sense—in case of ethnic mobilization within Ethiopia by another, major group—and in the absolute sense, by way of assis-

---

[24]For more background information, see Patrick Gilkes, *Conflict in Somalia and Ethiopia*, New York: New Discovery, 1994.

[25]But it should be noted that in the investigation of at least one of the bombings, the Al-Ithad Al-Islam might have been used as a convenient scapegoat by the police. "Explosions in Addis Ababa, Dire Dawa," *Ethiopian Review*, May–June 1997, p. 12.

tance from Somalia. The clan-based structure of the Somali provides effective channels of organization through which to funnel and use the resources. The clan structure also provides a means of legitimization for a potential Somali leader (assuming skillful leadership that could unify the various factions within the clans). Moreover, a moderate to high degree of closure provides readily identifiable grievances around which to mobilize the group. The potential of a windfall from natural resource exploration could fuel further the mobilization.

## ASSESSING THE STATE

### Accommodative Capability

How inclusive and responsive are Ethiopian political structures to the general population? The political institutions in Ethiopia are ostensibly democratic and inclusive, though in practice they have failed to fulfill their functions, as few opposition groups so far have chosen to participate formally in the electoral process. The view of the governing structure as essentially a sophisticated fig leaf for Tigrayan control has discouraged widespread participation. The goal of the opposition parties that do not agree with EPRDF policies is to "deny legitimacy" to the government.[26] The clearest example of group exclusion concerns the Oromo Liberation Front. The OLF has opted out of the formal political process since 1991 and the formation of the transitional government of Ethiopia. This is troublesome because the Oromo are the single largest ethnic group in Ethiopia. But if they want to participate in government, non-Tigrayan groups are required to accept a strong Tigrayan role in defining Ethiopia and its political agenda. Many organizations may not relish cooperation of this sort, since their constituents may interpret it as collaboration, with all the attendant negative repercussions.

Is there potential for change in Ethiopian political structures? Mechanisms for conflict resolution and institutional change are built into the Ethiopian constitution. According to it, questions of federal and regional laws that contradict the constitution are sent to the upper chamber of the parliament (the Federal Council) for resolution. Pro-

---

[26]Lyons, "Closing the Transition," (1996).

posed amendments to the constitution follow a more complicated process through the regional councils and then both national houses of parliament.[27] The constitution includes a process by which ethnic groups may secede from the Ethiopian state.[28] Because the constitution has been in force only since mid-1995, its provisions for conflict resolution remain to be tested. In other words, they exist in theory but have not yet been practiced.

In short, it is too early to assess whether the channels for institutional change are effective. Nominally, the political system is inclusive, and its decentralization aims at a high level of participation (from at least the major ethnic groups). But the inclusiveness for now is embedded in the larger context of Ethiopian recent history, that is, within the constraints of a completely new system of governing set up by an ethnically defined group that won a lengthy civil war. As such, the system has not managed to convert the skeptics, as most of the non-Tigrayan population appears to view it with a measure of distrust, a Tigrayan creation. As the system develops and matures, the opposition parties may decide to "legitimize" the political process with their participation. The election in the year 2000 will be key in determining the relative inclusiveness or exclusiveness of the regime.

In terms of prevailing norms of governance, neither the autocratic imperial Ethiopian government (pre-1974) nor the Derg (1974–1991) were tolerant of opposition. The Derg used widespread violence and repression to deal with any kind of dissent, real or imagined. In comparison with such a past, the post-1991 norms of governance have been remarkably tolerant and inclusive, even though there has been some retrenchment since 1994.[29] Initially, the EPRDF made a

---

[27]The Constitution of the Federal Democratic Republic of Ethiopia, 1995.

[28]Some have argued that the constitution reflects the very nature of the Tigray rise to power and its attempt to redefine the state as one comprising many distinct cultures; thus, the right to secede is a fundamental aspect of the TPLF view of Ethiopia. Assefaw Bariagaber, "The Politics of Cultural Pluralism in Ethiopia and Eritrea: Trajectories of Ethnicity and Constitutional Experiments," *Ethnic and Racial Studies*, Vol. 21, No. 6, November 1998, pp. 1056–1073.

[29]Since 1994 and the end of the transitional charter, the government's behavior toward groups that opposed or declined to participate in the political process has shown a trend toward less tolerance, with incidents of intimidation and force used against domestic opponents. The incidents include the arrest and imprisonment of journalists, a public war of words with opposition groups, and crackdowns on demon-

distinct attempt to include other groups, such as the OLF, in the governing coalition. It also allowed a free press to flourish and sought the political opinions of the populace (i.e., the constitutional "discussions" that occurred at the neighborhood level). Thus, the governing structure set up by the EPRDF has tried for inclusiveness in a way that showed a real break with the previous patterns of governance, even if the limited participation by non-Tigrayan groups in the governing process so far suggests their persistent distrust. In addition, the ethnically based federal structure shows the vitality of ethnically defined collectivist norms. Finally, for many Ethiopians, including members of the Oromo and many smaller groups, an Ethiopian identity is weak and secondary to their ethnic attachments.[30]

What is the level of cohesion among the ruling elites? Little information is available about the inner workings of the EPRDF government, but there are some signs of discord. The discord may center on policy differences rather than fundamental issues on the setup of the state. The very public firing in 1996 of the deputy prime minister and minister of defense, Tamirat Layne, may be an indication of a power struggle. Whether this was the elimination of one individual or an indication of deeper splits and factional power struggles within the ruling coalition is uncertain. Adding to suspicions of conflict within the party was a purging of OPDO party members in April 1997. Eighty OPDO members were subsequently placed in custody. This development suggests that cohesion among the EPRDF coalition partners is declining and that the EPRDF is willing to take action against partners who are not toeing the party line. The trend appears to be toward a more isolated Tigrayan-based governing coalition. Whether the trend means an increase or decrease in leadership cohesion is impossible to say at this time.

In conclusion, the accommodative capability of the Ethiopian state is difficult to assess in an unequivocal manner because of the ambiguities surrounding it. The institutional structure of the state seems

---

strations. In part at least, the more forceful government reaction is a response to the non-participation of other groups in the process. S. F. Joireman, "Opposition Politics in Ethiopia," September 1997; "Students Say No to Petition and Remain in Detention," *Addis Tribune* (Addis Ababa), April 4, 1997.

[30]See the Afterword to this chapter for potential changes to this situation.

viable for an inclusive and responsive system, but the legacy of a long period of interethnic tensions and distrust remains a practical obstacle to achieving the democratic potential of the institutional structure contained in the constitution. A move by the coalition toward less inclusiveness of other groups may, in fact, have the effect of strengthening the internal cohesion of the ruling group.

## Fiscal and Economic Capability

The fiscal health of Ethiopia shows serious strains, and the mainstay of the economy—agriculture—is subject to weather- and climate-induced disruptions. However, the macroeconomic indicators have stabilized and the economy has registered consistent growth since the end of the civil war (1.7 percent in 1993–1994 and 4.9 percent in 1994–1995). The state's economic targets through 1999 include 6 percent real GDP growth, limiting the external current account deficit to 9.2 percent of GDP, and maintaining low inflation. Although the goals are ambitious, they are realistic, assuming no major drought and/or worldwide recession.

The state has followed a consistent pattern of deficit spending. In the fiscal year 1994–1995, the Ethiopian budgetary deficit reached 3.8 percent of GDP. Ethiopia also has a heavy debt; the World Bank rates Ethiopia as severely indebted and "possibly stressed" regarding the sustainability of its debt burden.[31] In 1995, approximately 13.6 percent of Ethiopia's earnings from exports went to the payment of the interest and principal on debts.[32] Ethiopian gold reserves have remained steady, and its foreign exchange position has improved significantly since 1992. The state also has further loans it could call on from international lending institutions. See Table 5.7.

Widening the tax base and improving tax collection remain central objectives of the government. Customs duties and sales taxes remain important for revenue generation. Taxes on personal and business income and profits have risen substantially. Both trends imply an improved tax-collection system. There is some potential for the

---

[31]World Bank, *Global Development Finance*, Volume 1, Washington, D.C.: The World Bank, 1997, p. 43.

[32]Ibid., p. 216.

reappropriation of funds within the state budget. Although health and education have registered substantial growth since 1992 and strict defense expenditures have gone down to 13 percent of the current expenditures in the 1994–1995 fiscal year, there seems to be enough leeway for further shifting of funds on a short-term basis. See Table 5.8.

The wealth of the core constituency of the ruling elite does not differ at all from the wealth of the general population. Neither this group nor any other in Ethiopian society faces a disproportionate tax burden. Some accusations have surfaced that a disproportionate amount of development aid is directed toward the Tigray region, but since Tigray received little development assistance from the government between 1984 and 1991, even if this were to be the case it would

## Table 5.7

### Ethiopian National Bank Reserves (in US$ million)

|                            | 1993  | 1994  | 1995  |
|----------------------------|-------|-------|-------|
| Gold (national valuation)  | 11.4  | 11.4  | 11.4  |
| IMF special drawing rights | 0.3   | 0.4   | 0.3   |
| Reserve position in IMF    | 9.6   | 10.2  | 10.5  |
| Foreign exchange           | 445.9 | 533.6 | 760.8 |
| Total                      | 467.2 | 555.6 | 782.9 |

SOURCE: IMF, International Financial Statistics.

## Table 5.8

### Ethiopian Government Expenditures (in millions of birr)

|                                            | 1992–1993 | 1993–1994 | 1994–1995 |
|--------------------------------------------|-----------|-----------|-----------|
| Total                                      | 6,054     | 7,492     | 8,152     |
| Capital expenditure                        | 2,150     | 3,018     | 3,077     |
| Current expenditure                        | 3,904     | 4,474     | 5,075     |
| State administration, defense, and police  | 1,173     | 1,453     | 1,561     |
| Education and health                       | 583       | 742       | 936       |
| Interest and charges                       | 530       | 952       | 855       |

SOURCE: National Bank of Ethiopia

hardly change the wealth of Tigrayans vis-à-vis other Ethiopian ethnic groups.

In sum, in an absolute sense, the fiscal capacity of the Ethiopian state is weak.  But within the constraints of being one of the poorest countries in the world, Ethiopia's state policies of economic liberalization have borne some fruit.  Although the fiscal and economic resources and capabilities available to the government remain low, the situation is improving.

## Coercive Capability

Cabinet ministers, subordinate to the prime minister, control the police and the armed forces.  The police are under the control of the Justice Ministry.  The minister of defense (Teferra Walwa) is constitutionally required to be a civilian.[33]  In addition, Prime Minister Meles Zenawi has his own security team headed by TPLF Central Committee member Mulugeta Alemseged.  Rather than an instrument of control over portions of the population, the team is oriented more toward the protection of the top leadership.  Formally, there is constitutional provision for a state of emergency; at such a time, there is a restriction on "fundamental political and democratic rights."  The Council of Ministers can declare a state of emergency when there is danger to the "constitutional order" that cannot be brought under control by means of normal law enforcement.  The Chamber of People's Representatives must approve any declaration of a state of emergency, though under the conditions of near-complete EPRDF control of the legislature, this provision is not a major obstacle.  In practice, despite the formal rules restricting the use of force, extrajudicial killings and disregard for civil rights and due process within the country have grown in number.

Who serves in the apparati of violence?  Although the government has attempted to form a "multiethnic defense force" consisting of all of Ethiopia's ethnic groups, the armed forces continue to be based on the armed wing of the EPRDF (and especially the TPLF) that won the civil war.  A substantial number of personnel from the OLF were also absorbed initially into the single post–civil war Ethiopian mili-

---

[33]Article 95, Constitution of the Federal Democratic Republic of Ethiopia.

tary, though their continued presence is in doubt in view of the OLF's withdrawal of its support of the government in 1992.

Official statements claim that the military is now far more representative than it was in 1991, when it was predominantly Tigrayan with some Amhara participation.[34]   In line with the striving for greater inclusiveness, new recruits of all ethnic backgrounds go on to officer training if they are assessed as having promise.  According to a recent USIS report, in 1995–1996, 25,000 primarily Tigrayan soldiers were demobilized and replaced by recruits from other ethnic groups.[35] However, several thousand of the demobilized Tigrayan soldiers have become policemen.[36]  One can raise questions about how loyal Tigrayan police officers assigned to regions other than Tigray will be to the regional police hierarchy.

What norms exist with respect to the use of violence domestically? Every government in Ethiopian history has used the armed forces to crack down on domestic uprisings.[37]  Indeed, there is no precedent within the country for government restraint in its actions against citizens.   The EPRDF government, since 1991, has been less violent toward its citizens than any other in recent history, but accusations of extrajudicial killings and beatings in the south and torture within the prisons have surfaced.[38]  More recently, in May 1997, an alleged leader of the Ethiopian United Patriotic Front was killed in Addis Ababa under suspicious circumstances (with government spokesmen claiming he was "resisting arrest").[39]

Ethiopian culture, insofar as can be generalized among all its ethnic groups, has a relatively low threshold for violence.  Violence is often used as a means of political expression, and there is only limited

---

[34]"Effort to Build Multi-National Defense Force Successful," Ethiopian News Agency (Addis Ababa), July 23, 1996.

[35]USIS, (1996).

[36]"Looking Federal," *Africa Confidential,* September 22, 1995, p. 5.

[37]See Gebru Tareke, *Ethiopia: Power and Protest,* New York:  Cambridge University Press, 1991; Africa Watch, *Evil Days: Thirty Years of War and Famine in Ethiopia,* New York:  Human Rights Watch, 1991.

[38]Lyons, "Closing the Transition," 1996.

[39]Reuters African Highlights (Addis Ababa), May 8, 1997 (Internet version).

experience with ideas of human rights as enshrined by the United Nations.

Are the armed forces suitable for domestic use? The armed forces are largely light infantry, suitable for domestic use, especially in an environment of permissive norms toward violence. Most soldiers have extensive experience fighting in the Ethiopian countryside, gained during the civil war. The military could be employed in a range of activities, from forceful crowd control to counterinsurgency. Training for domestic use is likely to be rudimentary.

In conclusion, the state can readily resort to violence, and the apparati of violence have considerable capabilities against domestic opponents. Use of force in the domestic context is constitutionally allowable through the imposition of a state of emergency.

## STRATEGIC BARGAINING

This analysis of the potential mobilization of the Amhara, and to a lesser degree that of the Oromo and Somali groups, on the one hand, and the capabilities of the Ethiopian state to deal with any such mobilization, on the other, provides a means for considering the possible interactions between them based on the categories and matrices of the framework presented in Chapter Two. As in the two previous cases, the group and state types are categorized on the basis of their capacities, and the matrices provide a way to think about their interaction conceptually.

### Measuring the Group's Capacities

The leadership of the Amhara is confident. This confidence stems from the mythology of the group, including perceptions of manifest destiny, the view of the current government as unrightfully in place, and the strong rhetoric supporting these views emanating from the Amhara diaspora in diverse areas of the world. The leadership has not hesitated to take risks. The radical Amhara activists are united in their intent to overthrow the current government, though what type of regime they think should take its place, a predominantly Amhara state or a multiethnic state, is not yet clear. Thus, the assessment of leadership that a potential mobilized Amhara is likely to have is "strong."

With respect to resources, support for the Amhara is good, especially in the relative sense of the poor conditions in Ethiopia. The Amhara activists have an extensive support network overseas, and they draw their support within the country from the wealthiest and most educated people. At present the resource base is sufficient to support an urban-based movement against the government. The available support is well-suited to the goal of ousting the current regime from power. But it may not be enough if the objective is to establish a predominantly Amhara rather than multiethnic state. Thus, the assessment of resource support that a potential mobilized Amhara is likely to have is "good."

Regarding popular support for the Amhara mobilization, the activist Amhara cohort so far is primarily urban. The more numerous Amhara in the countryside have not been mobilized. Land reform may yet change the situation, but there is little evidence of it at the present. In addition, there are no readily identifiable other groups that seem willing or likely to support the Amhara. The Tigrayans are currently in control of the government and are, therefore, the enemy. The Oromo are extremely unlikely to form any sort of alliance with the Amhara, for many Oromo consider them former "oppressors." Thus, the assessment of popular support that a potential mobilized Amhara is likely to have is "weak."

Based on these assessments, the mobilized Amhara is judged to be a type D group. The capacities of such a group are as follows:

Accommodative: low;

Sustainment: high;

Cohesiveness: high.

## Measuring the State's Capacities

Concerning the leadership of the state, the EPRDF government built a guerrilla army and created a development arm and a well-constructed ideology before it won a civil war in 1991. After emerging as the victors in the war, the EPRDF then tried to build a multiethnic ruling coalition for Ethiopia, with EPRDF at the core. It also wrote a new constitution and embarked on development projects within the country. There is no question that the leaders of EPRDF are secure in their position and beliefs. They have attempted to make the system of governance inclusive to all Ethiopians but have not hesitated to use forceful means in implementing their policies. They have a clear vision of a more developed and prosperous Ethiopia, based on consociational ethnic relations and a market economy. Thus, the assessment of leadership is "strong."

In terms of its fiscal position, the state has engaged in deficit spending and has become deeply indebted. There is some room for reallocation of funds, within the overall constraints of a low state budget. If the economy continues to expand, there are more optimistic prospects for the future. But the agricultural base of the economy and a dependence on the export of primary products as the source of most foreign exchange puts in place a low ceiling for growth and revenue generation. Thus, the assessment of fiscal position is "weak."

Regarding the regime type of the state, this capacity is not as clear-cut as the others. The Ethiopian regime exhibits the characteristics of both types. Institutions exist for competitive elections and the replacement of government officials; however, these institutions have not been put to the test sufficiently to conclude without hesitation that the regime is inclusive. A largely free media exists, though some harassment and intimidation of the press is evident. There are limits on executive power, but there are also prevailing norms of intolerance. The tendencies will be easier to ascertain in another few years, especially after the elections in 2000. Yet it is clear that the current regime is much more inclusive than the previous regimes in recent Ethiopian history. Thus, on a close call, the regime is assessed as "inclusive."

Based on these assessments, the Ethiopian state is judged to be a type C state. The capacities of such a state are as follows:

Accommodative: high;

Sustainment: low;

Coercive: low.

Note that if the regime type is assessed as exclusive, then the coercive capacity is high, and the state type is G.

## Outcome of Bargaining and Preferences for Violence

Based on the matrix showing the preferences of the mobilized group, a type D group has the following preferences toward a type C state: (1) negotiate, (2) exploit, and (3) intimidate. Based on the matrix showing the preferences of the state toward a mobilized group, a type C state has the following preferences toward a type D group: (1) repress, (2) exploit, and (3) negotiate. Note that if the state type is G, due to an assessment of the regime as exclusive, the group's preferences change to exploit, negotiate, intimidate. The state's preferences remain the same: repress, exploit, negotiate.

Comparing group and state preferences leads to the conclusion that the potential for violence in the dyadic encounter between a primarily urban-based mobilized Amhara and an EPRDF (Tigrayan) Ethiopian state is high. The violence-prone outcome stems primarily from the state's preference for a strategy of repressing the challenger, with a hedging strategy of exploiting the weakness within the mobilized group. In practical terms, the latter may mean a forceful crackdown and intimidation of the urban Amhara and incentives to the rural Amhara to continue their acquiescence. Quite simply, the EPRDF has few resources with which to head off dissent, but it has a disciplined and cohesive leadership, a blueprint for the future, and control of substantial apparatus of violence, all of which make it determined to stay in power, by force if necessary. The preferred Amhara strategy is to negotiate, with a hedging strategy of exploitation or, failing that, perhaps even outright intimidation of the state. But the pattern of preferences in strategies clearly illustrates the weakness and power difference between the group and the state.

If the state is assessed as an exclusive regime type, then the group's first two preferences are reversed. A more forceful strategy of exploitation is the preferred strategy, for the state makes the situation of the Amhara worse and the group then needs to demonstrate its resolve.

The choices in the original matchup are interesting because they are the inverse of each other. The state starts from a position of strength and has incentives to keep its superior position. On the other hand, the Amhara start from a position of weakness and are wary of an escalation to violence because that will put the group at risk. The group has incentives to continue to negotiate (forcefully, through the exploit option, if need be) but not to resort to all-out force.

If the Amhara leadership were able to mobilize the rural Amhara, turning the group type to A, its choices against a still predominantly inclusive regime type (state type C) would stay the same. Moreover, the group would stand a real chance of having a larger access to power, because the state's preference would shift to negotiation as the preferred strategy. But if an Amhara shift to type A is accompanied by a state move toward a more exclusive regime type (state type G), then violence becomes the preferred strategy for both sides. Intelligence analysis needs to be on the alert for such an evolution in Ethiopia.

What is telling about the choice of strategies is that the state has little tolerance for any challenges from the Amhara, and the state's resource base supports a strategy that includes a quick recourse to force. The Amhara are weaker and, from a resource base perspective, their demands and tactics are likely to be moderate.

**The Oromo.**  With respect to the secondary scenarios of mobilization, the current situation of the Oromo is characterized by weak leadership, weak resource support (predominantly poor and rural), and broad popular support (largest single ethnic group, with appeal potential beyond the Oromo), which make it group type E. The group's preferences against state type C (inclusive regime) are exploit, negotiate, intimidate. The state's preferences against the group are exploit, negotiate, repress. If the state type is assessed as G, then the group's preferences are negotiate, exploit, surrender, and the state's preferences become exploit, negotiate, repress. In neither of the four sets of preferences is violent outcome either a first or second preferred strategy. According to the framework, there is little potential for Oromo-centered violence.

If the Oromo move toward a strong leadership (group type C), then the group's preferences against state type C (inclusive regime) are negotiate, exploit, intimidate. The state's preferences are negotiate, exploit, repress. In other words, even if the Oromo mobilize with a strong leadership, there is still little potential for violence if the state stays inclusive or moves further toward inclusion of other groups. However, the Oromo with a strong leadership (group type C) facing a state type G (exclusive regime) have very different preferences: intimidate, exploit, negotiate. The state's preferences then become repress, exploit, negotiate. In other words, the primary strategic preferences for both the state and the group are for the use of violence.

**The Somali.** The current situation of the Somali is characterized by strong leadership, weak resource support, and weak popular support (small overall numbers, limited appeal to other groups), which make it group type G. The group's preferences against state type C (inclusive regime) are negotiate, exploit, intimidate. The state's preferences against the group are repress, exploit, negotiate. If the state is assessed as G, then the group's preferences are exploit, intimidate, negotiate. The state's preferences remain repress, exploit, negotiate. In short, the state has a preferred strategy of violence toward the group. The group's preference for violence increases as the state turns toward greater exclusiveness. According to the framework, there is good potential for violence to accompany any Somali mobilization.

If the Somali move toward greater resource support (group type D), perhaps as a result of assistance from a reconstituted Somalia, then the group's preferences against state type C (inclusive regime) are negotiate, exploit, intimidate. The state's preferences are repress, exploit, negotiate. Against a state defined as an exclusive regime (state type G), the group's preferences are exploit, negotiate, intimidate. The state's preferences remain repress, exploit, negotiate. In other words, the state has a preferred strategy to turn to violence to counter a Somali mobilization in conditions of both limited or strong resource support from Somalia. Given the shape of the state's responses to Somali activism so far, the assessment is on the mark.

## Revisiting the Evaluation

There is much to be said for the portrayal of the type of ethnic tensions between the Amhara and the Tigrayan-dominated state as expressed in the terms above. The Amhara have followed the identified strategic preference for negotiation. The preference does not mean that some activists within the group are not preparing for more forceful action, and factions of the Amhara, both in Ethiopia and abroad, advocate strong measures.[40]  But so far, the Amhara have "negotiated" through implied threats and a more pronounced mobilization effort.  The Amhara have launched neither an insurgency nor even a terrorist campaign; they have stayed largely within the bounds of peaceful, if sometimes attention-getting, methods.

The state has shown little tolerance for any challenger.  It has acted to repress vocal dissenters, and its potential for violent measures seems to be understood by all the major actors in Ethiopia.  Whether the state moves toward being more or less inclusive during the next few years is crucial.  If it moves toward greater inclusivity, then the strategic preferences of any challenging groups will also change, usually toward more peaceful methods of taking on the state.

The Oromo and Somali relationship with the state also falls within the bounds of the conceptual framework of the model.  The main Oromo groups disagree on the extent of participation in the government, with the differences primarily based on greater sense of separation or inclusion of the Oromo in Ethiopia.  Negotiation strategies predominate, as borne out in reality.  Similarly, the Somali relationship is captured in the preference of the state for repression no matter which of the two main paths the Somali mobilization takes.  The heavy-handed state control over the Somali (emphasizing the role of the police), indicates the accuracy of the framework.

For the purposes of tracking the evolution of interethnic relations in Ethiopia, any appearance of the following trends needs special monitoring by intelligence analysts:

- A clear move by the EPRDF away from the attempt at inclusion of other ethnic groups in the governance of the country and toward a more narrow emphasis on the Tigray.  Specific indicators might include a dwindling influence of non-Tigrayan parties (the Amhara National Democratic Movement [ANDM] and the Oromo People's Democratic Organization [OPDO]) within the EPRDF, greater prominence of Tigray mythology in EPRDF's

---

[40]Taye Wolde Semiat and the Ethiopian Patriotic Front, now also the Ethiopian United Front.

statements, or even the creation of a Tigray-centric vision of Ethiopia as a guiding plan for the future evolution of the country. In terms of the model, this would represent a move toward state type G, one of the most violence-prone state types.

- A significantly higher level of openness to and acceptance of urban Amhara appeals by the rural Amhara. Specific indicators might include the loss of support for the ANDM in the rural Amhara areas, a changed pattern of activity of the Amhara militant groups, and changes in philosophy and pronouncements of the urban Amhara activists to include concerns central to the rural Amhara. In terms of the model, this would represent a move toward group type A, one of the least violence-prone group types.

- The evolution of ties among the Oromo organizations toward greater cooperation or even unity. The mobilizing message among the united and mobilized Oromo would be important: Is it secessionist, or does it emphasize inclusion in the current Ethiopian structures? In terms of the model, the crucial variable is the type of state such an Oromo group would face. If it were an exclusive regime type, then the matchup would be highly violence-prone.

- Concerning the role of Somalia in the Ogaden, the state's heavy-handed attitude toward the Somali is unlikely to change, but in case of a revitalized Somalia emerging in the future, the intra-Ethiopian Somali-centered tension may attain again the inter-state conflict dimension that it has had several times in the past.

- The use of Islamic rhetoric and Islamic appeals, especially by the Oromo and the Somalis. If, in response, the EPRDF turns toward greater exclusion of these major non-Tigray groups, then, in terms of the model, the likelihood of violence will be enhanced.

- The ongoing process of regionalization and, specifically, the treatment of minority groups within each ethnic state. For example, perceived mistreatment of the Amhara minority in the Oromo state (in terms of land allocation or equal administration of justice) would escalate the grievances of and potential for the mobilization of the Amhara.

Thinking about the dynamics of ethnic relations and potential for strife in Ethiopia, the framework presented above makes apparent the following intelligence needs and information gaps:

- Tracking the rise of militant activists among the urban Amhara and any evidence that points to their increased preparations for armed actions against the government.

- Tracking the appeal of the urban Amhara mobilizational appeals among the rural Amhara, including the interest by urban Amhara activists in the rural Amhara, visits, common organizations, and organizational channels to further the mobilization process.

- Following closely the appeal of EPRDF to non-Tigrayan groups, including signs of willingness by the Amhara, Oromo, and Somali to participate in the governance of the country and the popularity of major non-Tigrayan EPRDF political figures among their ethnic cohorts.

- Following closely the appeal of anti-EPRDF, ethnically based political groupings in the Amhara, Oromo, and Somali regions.

- Analytical scrutiny of the 1994 census data, including taking into account the probable errors of under- and overcounting, the identification of minority groups within regions (for example, Amhara living in Oromo areas), their projected rates of growth, and potential for increased ethnic competition at substate (regional) levels.

- Additional information on the police, including the principles used (in theory and in practice) for selection and promotion within the police force, the ethnic composition of the regional police forces, and the relationship between the police and the armed forces at the local level.

- Additional information on the armed forces, including the changing ethnic composition of the officer corps and the actual criteria for promotion, the principles behind stationing and assignment of troops in the regions, as well as indications of the overall prestige of the armed forces among the various ethnic groups.

Close attention to these issues may enable an early identification and warning of impending ethnic strife in Ethiopia.

## FINAL OBSERVATIONS

The framework presented here examines the situation in Ethiopia through a series of dyads, pitting a specific mobilized ethnic group against the Tigrayan-dominated Ethiopian state. As such, the model does not capture the various intergroup relations that may be influential in the equation. For example, Amhara mobilization may spur greater Oromo support for the Tigrayan-led government, or, depending on the course the mobilization takes, it could spur the Oromo to undertake their own mobilization against the Tigray. In other words, a weakened Tigray-led Ethiopian state may come under challenges from both the Amhara and the Oromo, with the Amhara-Oromo relations ranging anywhere from tacit alliance to mutual hostility (surpassing even their hostility toward the Tigray-led government). On the other hand, the ethnic mobilization of the Amhara against the Ethiopian state is interesting in the sense that it stipulates the mobilization of an ethnic group to restore the privileges the group once had. The two secondary scenarios, focusing on the Oromo and the Somali, add further depth to the analysis.

Critics may believe that the lack of a multiactor perspective to the model is a serious flaw that detracts from its usefulness. The lack may be a necessary shortcoming to make the model user-friendly, but it need not be a fatal flaw. The framework provides a tool for thinking abstractly about the mobilization patterns and challenges to the ruling state authorities. It is meant to provide insight into the strategic preferences available to the main actors under certain "what-if" conditions. As such, the model aims to be parsimonious, and its narrowing of the crucial questions to the main dyad illuminates the logically derived preferences of both actors without unwieldy mathematical equations. The way to capture the influence of other parties on the crucial dyad is to include their impact when calculating the state's and the group's capabilities. For example, in accounting for the Oromo factor in the Tigray-led state against Amhara dyad, the limited appeal to the Oromo of the Amhara mobilization can be captured in the assessment of the sustainment and cohesiveness capacity of the Amhara. Similarly, a potential Oromo disinclination to defend the Tigrayan-led state against a challenging Amhara can be captured in the assessment of all three state capacities.

Since the model was applied to a prospective case, its accuracy in anticipating the likely strategies that the challenging groups and the state are likely to follow is, as yet, unverifiable empirically. But the purpose of applying the model to a prospective case was to provide a series of hypothetical but likely situations and determine the evolutionary paths for both the state and the group that would result in the most violence-prone confrontations. As such, the model proved useful in illustrating the "dangerous dyads."

Finally, the model was designed to structure intelligence analysis, trace the logical connections in the evolution of tensions along interethnic lines, and provide a framework for thinking of potential outcomes to hypothetical situations. Outlining specific trends for analysts to track—on the basis of their potential for violence—is the goal of the exercise. The model is a tool toward that end, as it provides the all-important linkage between goals, level of resources, and the resulting strategies and choices open to the state and the group.

## AFTERWORD

In May 1998, war broke out between Eritrea and Ethiopia. The war was ostensibly over a small patch of land on the border of Eritrea and the Ethiopian state of Tigray. However, the armed conflict was precipitated by economic friction between the two countries that began when Eritrea (officially independent of Ethiopia from 1993) launched its own currency, the nafka, and demanded that all trade and exchange between the two countries be conducted in hard currency. This was a particularly burdensome request for Ethiopia, since it uses the Eritrean port of Assab to export many of its tradable goods. Fighting between the two countries continued for a few weeks before both the UN and the Organization for African Unity began to try to mediate some sort of peace agreement. These negotiations broke down, and conflict erupted again in February 1999.

These events have reduced significantly the potential for ethnic conflict within Ethiopia. The ethnic divisions that were so pronounced before the fighting broke out have receded with the appearance of a new and external enemy. Ethnic divisions within Ethiopia, and ethnic divisions between Ethiopian groups outside of the country, have become muted as nationalist sentiment is generated by the war. It is unclear whether or not the war will have a long-term positive effect on the possibility of ethnic conflict in Ethiopia. The length of the war, its outcome, and subsequent government policies will all play a role in determining the eventual result. The length in particular will be important. As the war drags on, the elections of 2000 may be far less contentious than previous elections. But when the war ends, the country could be back to ethnic politics as usual and the scenarios documented herein will be as relevant as ever.

# ANNEX:  DEMOGRAPHIC CHARACTERISTICS OF ETHIOPIA IN 1997–1998

The following information about population characteristics is based upon the situation in Ethiopia at the beginning of 1998.  The information presented here is the basic reference for the analysis in this chapter.

**Name:**  Federal Democratic Republic of Ethiopia.

**Capital:**  Addis Ababa.

**Nature of government:**  Nominally parliamentary democratic system in a federal state with a strong federal prime minister and substantial local powers held by the regions.  The federal parliament consists of two chambers:  the Council of People's Representatives (the lower chamber) and the Council of the Federation (the upper chamber).  The lower chamber has lawmaking functions and consists of 548 directly elected deputies.  The upper chamber is responsible for reviewing constitutional and regional issues and has 117 deputies chosen by assemblies in each region.  Judges of the Supreme Court are elected by the legislature.

**Organization of the state:**  A federal state, with nine ethnically based administrative regions and two cities with regional status:  the federal capital region of Addis Ababa, and Harar.  The regions are Afar, Amhara, Benshangul/Gumaz, Gambela, Oromia, Somalia, Southern People's Region, Nationalities and Peoples, and Tigray.  Two of the regions (Southern People's Region and Nationalities and Peoples) contain several small ethnic groups and are defined as territorially defined amalgamations of ethnic groups.

**Date of constitution:**  December 1994 (in effect since August 1995).

**Population:**  58.7 million (estimate, July 1997).

**Major ethnic groups:**  Amhara, Oromo, Tigrayan.  Dozens of ethnic groups exist in Ethiopia; classification of the population along ethnic lines is difficult because of the variety of determinants of ethnicity, such as language, religion, region, and historical experience.  Some ethnic groups may share a language and/or religion.

**Languages:**  Major languages are Amharic (official language), Tigrinya, Oromiffa, Guaraginga, Somali, Arabic, English.  More than seventy languages are spoken in Ethiopia, most of them belonging to three families of the Afro-Asiatic language group:  Semitic (Ethio-Semitic and Arabic branches), Cushitic, and Omotic.  A small portion of the population (approximately 2 percent) are native speakers of language families belonging to the Nilo-Saharan language group: East Sudanic, Koman, Berta, and Kunema.  The principal languages within the Ethio-Semitic group are Amharic and Tigrinya.  Of the various Cushitic languages (subdivided into Highland and Lowland East Cushitic, Central Cushitic, and Northern Cushitic), Orominga and Somali are the most numerous.  Weleyta are the most numerous speakers of the Omotic language group.

**Religions:**  Islam, Orthodox Christianity, animism. (Catholicism and Protestantism also present.)

**Population statistics:**  Because of the difficulty in obtaining accurate population statistics, the data presented here should be interpreted as approximations.  The sometimes wide variance in estimates means that percentages of the total population—according to religious or linguistic differences—are more significant than the absolute figures.

Some Oromo factions believe the government grossly underestimates the Oromo population and land claims.  They claim traditional lands north of Addis Ababa and south to the Kenyan border.  These factions are correct insofar as particular "Oromo" groups have been classified by their subgroup (such as the Borana, who are Oromiffa speakers and considered part of the larger Oromo people, but whom the government defines as Borana).  However, not surprisingly, the Oromo idea of a "Greater Oromia" suffers from its own shortcomings.  For example, some Oromos claim that Addis Ababa, Debre Berhan, and Tegulet and Bulga are traditional Oromo lands, overlooking the fact that non-Oromo indigenous populations in some of these areas exceed 50 percent.

The Christian and Muslim populations of Ethiopia are generally mixed, but we can make some cursory generalizations about their regional distribution.  The population to the north of Addis Ababa, on the highland plateau, is predominantly Orthodox Christian.  The population to the south of Addis Ababa in the Rift Valley and east tends to be Muslim, and people west of the Rift Valley are Christian. Animist practices are found among the pastoralists inhabiting the far south of the country and scattered throughout the south.  Amhara

### Table 5.9

**Population of Ethiopia by Ethnicity**

| Ethnic Group | Percent of Population, CIA World Factbook | Percent of Population, Library of Congress |
|---|---|---|
| Oromo | 40[a] | 40[a] |
| Amhara | 32 | 30 |
| Tigray | 9 | 15 |
| Sidamo | 6 | NA |
| Shankella | 6 | NA |
| Somali | 4 | NA |
| Afar | 2 | NA |
| Gurage | 1 | NA |
| Others | 1 | 15 |

[a]Oromo groups estimate their numbers at 50–60 percent of the total population.  Feyise Demi, "The Oromo Population and the Politics of Numbers in Ethiopia," *The Oromo Commentary,* Vol. 6, No. 1, 1996.

### Table 5.10

**Religion Statistics**

| Religion | Percent of Population, CIA World Factbook | Percent of Population, Library of Congress |
|---|---|---|
| Orthodox Christian | 35–40 | 50 |
| Muslim | 45–50 | 40 |
| Animist | 12 | NA |
| Protestant/Catholic | NA | 2 |
| Other | 3–8 | NA |

and Tigrinya speakers predominate in northern Ethiopia and in the cities. Oromo speakers can be found throughout the south and west of Ethiopia, with a large concentration in Addis Ababa.

### Table 5.11

### Population of Ethiopia by Language Group

| Ethnic Group | Speakers (millions) | Percent of Population |
|---|---|---|
| Amhara | 15 | 28.5 |
| Oromo | 14 | 26.6 |
| Tigray | 4 | 7.6 |
| Weleyta | 2 | 3.8 |
| Somali | 2 | 3.8 |
| Gurage | 1.5 | 2.9 |
| Hadiya | 1 | 1.9 |
| Kembata | 1 | 1.9 |
| Afar | 0.5 | 1.0 |
| Harari (Aderi) | 0.03 | 0.06 |

SOURCE: *Ethnologue*, 13th Edition, Barbara F. Grimes (ed.), Summer Institute of Linguistics, Inc., 1996.

Chapter Six

# THE SAUDI ARABIAN PROSPECTIVE CASE

## Graham Fuller and Thomas S. Szayna

## INTRODUCTION

This chapter offers a final illustration of the use of the "process" model for anticipating prospectively the incidence of ethnic conflict by applying it to the potential case of the emergence of ethnically based violence in Saudi Arabia. Similar in structure to the Ethiopian case presented in the previous chapter, this case affords an opportunity to use the model to examine the potential grievances the Shi'a might have against the Saudi state structures dominated by the Saudi monarchy and the likely path that a mobilization of the Shi'a might take. The chapter examines the case of Saudi Arabia from the perspective of what an intelligence analyst might conclude about the propensity of Saudi Arabia toward ethnic violence were she to use the "process" model. Data available as of 1997–1998 were used to conduct this analysis.

Just as in the case of Ethiopia, the choice of Saudi Arabia as a case study to apply the model is meant in no way to suggest that Saudi Arabia is somehow predetermined to slide into ethnic tensions or strife. The choice was made on the basis of geographical diversity (as other case studies examine different regions of the world) and the country's regional importance.

Analysts familiar with Saudi Arabia may find this an unusual choice for modeling potential ethnic conflict, in view of the dominant perception of ethnic homogeneity among Saudi Arabians. Nevertheless, the Saudi Shi'a do fit our definition of an ethnic group. The Shi'a are a distinct group within the larger milieu of the Arab and monolinguistic population of Saudi Arabia. As defined in Chapter Two,

> ethnicity refers to the idea of shared group affinity and a sense of belonging that is based on a myth of collective ancestry and a notion of distinctiveness. The group in question must be larger than a kinship group, but the myth-engendered sense of belonging to the group stems from constructed bonds that have similarities to kinship . . . . The constructed bonds of ethnicity may stem from any number of distinguishing cultural or physical characteristics, such as common language, religion, or regional differentiation.

The common "marker" in the case of the Shi'a is religion, and on that basis the Shi'a have been stigmatized socially (by the monarchy and much of the Sunni population of Saudi Arabia) and systematically excluded from Saudi Arabia's political system. The Shi'a inhabit a geographically compact region of Saudi Arabia, and, based on the ascriptive understanding of religious affiliation and ethnicity that is dominant in the region, they have a myth of separate ancestry. In addition, evidence points to ongoing attempts to mobilize the Saudi Shi'a on a group basis.

As in any society, there are numerous other substate groups inhabiting Saudi Arabia, differentiated primarily on regional, tribal, and clan bases. Those groups may be important for intelligence analysts focusing on the potential for political conflict in Saudi Arabia. But it would be a stretch to consider them "ethnic" groups by our definition. The tribal and regional groups tend to be based on kinship groups (and have a distinct sense of common ancestry) and they may have elements of regional differentiation. But their sense of distinctiveness is not yet pronounced enough to warrant their treatment as ethnic groups to the same extent as the Shi'a.

As with the Ethiopian case, the authors emphasize that the model, at this point of its development, is intended to be more suggestive than predictive. Neither the choice of the Shi'a nor the potential outcomes examined should be taken to imply that the group will mobilize against the monarchy or that Saudi Arabia is somehow predetermined to slide into strife.

Organized in a fashion similar to that of the previous three chapters, four sections follow this introduction. The first section examines the structure of closure and provides an analysis of which ethnic groups appear to be privileged and which seem underrepresented in structures of power (the demographic characteristics of Saudi Arabia in

1997–1998, on which the analysis is based, are appended at the end of the chapter).  Then it examines the strengths and weaknesses of the challenging group—the Shi'a—looking at its potential mobilization process.  A brief discussion of secondary scenarios follows.  The second section looks at the capabilities the state—Saudi Arabia—might bring to bear in dealing with the challenging groups.  The third section examines the strategic choices, arrived at on the basis of the assessments in earlier sections, that the state and the group are likely to pursue vis-à-vis each other, given their resource base.  Finally, a few observations conclude the chapter.

## ASSESSING THE POTENTIAL FOR STRIFE

### Closure in the Political and Security Realms

In terms of closure in the political realm, the current political system in Saudi Arabia dates to the conquest of much of the Arabian peninsula in the 1920s and the establishment of the Kingdom of Saudi Arabia in 1932 by Ibn Sa'ud (of the House of Sa'ud), whose main power base was in the Najd (central portion of the kingdom).  All subsequent rulers have been the sons of Ibn Sa'ud, and the kingdom remains an absolute monarchy.  The royal family dominates all political life in the country to an extent that has few parallels in the contemporary world.  Rule by the royal family remains personalized and idiosyncratic rather than formalized in any substantive sense.  Informal customs and judgments that assess the level of "trust" and loyalty toward the regime on tribal and regional bases form the dominant criteria for closure in the political realm.  The royal family has a tight hold on all top positions of power.  Individuals associated with the tribes from central Najd are seen as most loyal.

The highest-ranking political authorities in Saudi Arabia belong to the royal family of Al-Sa'ud, which has virtually absolute power within the kingdom.[1]  The royal family (now numbering over 5,000 males) permeates all areas of government of any sensitivity and ensures that other posts are manned by loyalists.  King Fahd is the head

---

[1]Indeed, the name of the kingdom stems from the name of the royal family.  For this reason, dissidents and opponents of the Saudi family often refuse to use the designation "Saudi" for the country, referring to it only as "Arabia."

of state (he acceded to the throne in 1982). His designated successor is the Crown Prince (the king's half-brother). The king is also the prime minister. The top individuals in the Council of Ministers are all part of the royal family, all of the same paternal line, and many are of the same specific maternal lineage. Skilled technocrats not of royal family background occupy the key "technical" ministries: finance and national economy, petroleum and mineral resources, planning, industry and electricity. They are appointed by the king and serve entirely at the pleasure of the royal family. The "technical" ministers change frequently (sweeping changes took place in 1995). This is in contrast to the royal family ministers, who rarely change. Recent cabinet changes have further strengthened the hand of the Sudayri clan which dominates the royal family. See Table 6.1.

Stemming from royal decrees issued in 1992, a Consultative Council (Majlis al-Shura) was set up in December 1993. The king appoints its 90 (initially 60) members.[2] The council has an advisory function and can be dismissed at will. Its members are generally distinguished and well-educated members of Saudi society.[3] Further evolution of the Majlis is worth monitoring, though the body is unlikely to be allowed to challenge government policy except in the most technical of realms. A few interesting exceptions aside, the king has

### Table 6.1

### Top Saudi Arabian Political Officials

| Position | Name | Royalty |
|---|---|---|
| Prime Minister | King Fahd ibn Abd al-Aziz as-Sa'ud | House of Sa'ud |
| First Deputy Prime Minister, National Guard Commander | (Crown) Prince Abdullah ibn Abd al-Aziz as-Sa'ud | House of Sa'ud |
| Second Deputy Prime Minister, Minister of Defense | Prince Sultan ibn Abd al-Aziz as-Sa'ud | House of Sa'ud |

---

[2]See also "Saudi Arabia Expanding Membership in Its Consultative Council by One-Third," in *Mideast Mirror,* July 3, 1997.

[3]See profile of government officials and council members, available from Saudi government Internet sources, *http://www.saudinf.com.*

avoided the appointment to the council of independent-minded individuals who might criticize the regime.[4]

Saudi Arabia has no judicial branch of government. It has a body of religious scholars (the 18-member Council of Ulema) who offer judgment on the Islamic legitimacy of policy, based on Islamic law. Many decades ago the Ulema spoke out (privately) with some independence on issues directly related to religious policies, but over the years it has become more generally subservient to the king in all matters. It still may express private concerns to the king on certain issues. The members of the judiciary are traditionally from the Al al-Shaykh family, which has historically been intimately linked—including via intermarriage—to the ruling al-Sa'uds, thereby providing the ideological/religious foundation to the al-Sa'uds' secular rule. The current minister of justice is a member of the Ulema.

Saudi Arabia's provinces have no independent executive (only administrative) authority. Almost all governors of the provinces are members of the royal family. Particularly prominent is Prince Salman, a full brother to the king, who is governor of Riyadh and a major figure within the ruling family.

Little systematic information is available about director-level state employees. Anecdotal evidence suggests that, over time, these positions have been filled with technocrats, often with good foreign education credentials, but selection also seems to depend on loyalty to the regime.

In terms of closure in the security realm, the royal family controls all of the top security positions. The king is the ultimate authority in Saudi Arabia and is the commander in chief of the armed forces as well as the top boss for all state bodies. The main security organizations are the armed forces, the National Guard, the intelligence organizations, and the police. Ministerial-rank personnel head these organizations. The National Guard represents the country's premier "security" organization. It provides the bulwark of protection to the regime, in that it deliberately constitutes a balancing force against

---

[4]See "Islamists Sympathetic to Dissident Sheikhs Named to Saudi Shoura Council," *Mideast Mirror*, July 8, 1997.

the regular armed forces.  Unlike the military, the National Guard is stationed in the major cities and is positioned to block any antiregime action by the regular military.

Close relatives of the king are the highest-ranking individuals in charge of the specific bodies that deal with internal and external security.  See Table 6.2.  At the director level, the officer corps of the Saudi armed forces appears to be drawn almost exclusively from the royal family and those close to it.  The National Guard is even more tightly controlled by the royal family.  Its members are drawn from loyal tribes from the Najdi heartland, all carefully balanced off against one another.  All of the operations and intelligence direc-torates are headed by junior members of the royal family (who by-pass the chain of command to go directly to the minister of defense on operational decisions).[5]  Less is known about the police.  Appar-ently, many officers are drawn from tribes loyal to the royal family, primarily in the Najd.  However, it is difficult to generalize, as some foreign nationals are employed as officers.

**Assessment of closure in the political and security realms.**  Closure in Saudi Arabia is based, first of all, on ties to the royal family, and then on tribal and regional bases (because of assumptions about their loyalty).  The royal family dominates in an absolute sense the

### Table 6.2

### Top Saudi Arabian Security Officials

| Position | Name | Royalty |
| --- | --- | --- |
| Commander-in-Chief | King Fahd ibn Abd al-Aziz as-Sa'ud | House of Sa'ud |
| National Guard Commander | (Crown) Prince Abdullah ibn Abd al-Aziz as-Sa'ud | House of Sa'ud |
| Minister of Defense | Prince Sultan ibn Abd al-Aziz as-Sa'ud | House of Sa'ud |
| Minister of Foreign Affairs | Prince Sa'ud al-Faisal as-Sa'ud | House of Sa'ud |
| Minister of Interior | Prince Nayef ibn Abd al-Aziz as-Sa'ud | House of Sa'ud |
| Director of Foreign Intelligence | Prince Turki Bin Faisal as-Sa'ud | House of Sa'ud |

[5]Said K. Aburish, *The House of Saud,* New York:  St. Martins Griffin, 1996, p. 190.

upper levels of political leadership.  Other Najdis and technocrats (of various tribal affiliations but all of known loyalty) also occupy ministerial and high-level posts.  However, there is a qualitative difference in the level of power they have, as belonging to the royal family *de facto* carries far more weight than having a specific position (due to informal but dominant ways of going around formal channels of authority).  The Hijaz region is represented in the government far less than the Najd.  The Shi'a seem almost completely left out of the leadership levels.

Similarly, the royal family dominates completely at the highest levels of security.  Reliable information on the tribal and regional composition of the middle and upper ranks of Saudi security organizations is not readily available, but anecdotal evidence suggests that the most trustworthy from the royal family and Najdi loyalists predominate.

Are there provisions for dynamic change within the Saudi Arabian political system presented above?  Although there are no formal restrictions to positions of political influence, that means little in a state where formal legal issues carry little weight and informal but widely understood customs predominate.  Informal restrictions are high.  Instinctive "trust" in general terms seems to reside more on a regional/tribal basis than it does at the ideological level; for example, even outspoken Najdis may be seen as inherently more trustworthy and committed to the regime than possibly "loyalist" Hijazis.  It is impossible to verify the accuracy of such beliefs (that is, the extent to which Najdis really are loyal to the regime).  With increasing modernization, the accuracy of the association of automatic loyalty along a regional/tribal basis may be decreasing.

A formal process of change of the elites at the highest levels is lacking, and a broad range of restrictions acts to maintain the status quo.  The monarchy shows no intention of relinquishing power or even allowing for the possibility.  There are no elections, hence no voting rights.  There is no formal process or mechanism for legitimization of political elites except at the wish of the royal family.  Political office may be held only by royal appointment.  Suppression—often brutal—of overt dissent is a part of the governing process.  Political institutions in the formal sense remain in their infancy.  Power is not institutionalized and remains idiosyncratic.  The king has stated publicly and often that democracy is not an "appropriate" form of government for Saudi Arabia.

The recently established Consultative Council represents virtually the only possible instrument of peaceful change, but the body has a long way to evolve to achieve such a role, as its members are appointed by the king. Further, the body can be dissolved at will by the king. In legal terms, the idea of a legislative body contradicts the religious tenets of Wahhabi Islam (and other conservative interpretations), in which man-made legislation is seen in contravention of "God's law," or Shari'a law. Thus even a proposal to create a legislative body would be dangerous to the regime's legitimacy. Alternatively, the regime could move to permit election of members to provincial consultative councils, but such a move would establish a dangerous precedent and is unlikely.

If anything, the trends over the past few decades have made peaceful change more difficult. In the pre–King Fahd days, the government was smaller and the massive oil revenues had just started. Although the royal family dominated at the top, it was in a close alliance with the religious establishment, and it engaged in tribal alliances as well as with the military and the merchant, professional, and business classes.[6] That dynamic is no longer in place, as the religious class is uncomfortable with the corruption and self-indulgent lifestyles of the royals. Diminishing oil revenues and a growing royal family have damaged ties with the merchant and business classes because of the royal family's efforts to skim off percentages of all business deals. In addition, there is a greater scramble for top jobs from better-educated royals, squeezing the traditional professional class.

The distinctions based on presumed trust and assumed loyalty toward the royal family affect the Shi'a the most, though other groups (tribal/regional distinctions) also face these barriers to a varying extent. According to Shi'ite activists, Shi'a are denied access to the following positions: government minister, deputy minister, judge, officer in the army, navy, or air force, secret police, public company director, newspaper editor, education superintendent, border customs, and any position in the following ministries: royal

---

[6]See "Forecasts," *The Monthly Monitor of Trends and Developments in the Muslim World,* London, June 1995, pp. 2–3.

palace, the National Guard, Foreign Ministry, Justice Ministry, Hajj (Pilgrimage) Ministry, and Ministry of Islamic Affairs.[7] There are rare exceptions: two Shi'a attained a senior government position (though none at the cabinet level). Two of 90 members of the enlarged Consultative Council are Shi'a.

## Closure in the Economic Realm

There is little reliable data available on the economic sector in Saudi Arabia, and data at the subnational level are especially difficult to find. Petroleum and natural gas extraction (and associated industries) are the primary areas of economic activity, accounting for 40–45 percent of the gross domestic product and generating almost all (about 90 percent) of the kingdom's export income. Data on employment patterns are not available (and estimates are difficult because of the employment of many nonnationals in the industry and services), though the economically active population is approximately 6 million (with estimates of the share of agriculture at 1 million). The general pattern that seems readily evident is that income distribution is skewed substantially, with the royal family in a class by itself and then the rest of the population.

At the elite level, the royal family is clearly advantaged in its access to income-generating institutions. According to an opposition estimate, an ordinary prince with ten children and two wives (not unusual) received $260,000 a month, while top princes with senior positions in the government had income as much as $100 million per year.[8] Because of their access to levers of government and means of regulation, informal or formal payments to royal family members throughout the government are the norm. For example, anecdotal data suggest that members of the royal family increasingly tend to play a role in the commercial life of the kingdom as "silent partners," demanding a cut of most business deals (especially foreign) as "facilitation fees."[9] Moreover, access to wealth by nonroyals is pos-

---

[7]Based on author Fuller's interviews with Shi'a in exile, 1997.

[8]Aburish, p. 68.

[9]For more detailed information on the privileges of the royal family and its pervasive interference in business life, see Charles J. Hanley, "House of Saud," May 3, 1997, Associated Press.

sible only through royal largesse. The current situation is in contrast to the state of affairs in the 1950s, when wealthy Hijazi merchants helped bankroll the royal family as partial partners in the national undertaking.

Although explicit data on wealth distribution at the general population level are not available, the falling price of oil since the 1980s has led to a drastic decline in real per-capita income (from $14,000 a year to $4,000 since 1982, by one estimate).[10] World Bank estimates that during 1985–1995, GNP per capita declined in real terms, at an average rate of 1.9 percent (while population increased by an annual average of 4.3 percent in the same period). The drop has affected all areas of the kingdom, though not necessarily to the same extent. Investment patterns favor the Najd and the petrochemical industry in the Eastern Province (the Shi'a region), since the latter is the key oil-producing region. While industry in the Eastern Province has some beneficial effects on the local economy, employment patterns do not favor the Shi'a. Anecdotal evidence suggests that the Shi'a as a group have the lowest standard of living in the country.

**Assessment of closure in the economic realm.** The dearth of reliable data makes conclusions difficult. However, it is clear that a huge gap in income exists between the small elite, composed primarily of the royal family, and the rest of the population. The income gap appears to be growing, with the elite becoming even more rich while the general population is competing more intensely for a share of an increasingly smaller pie (due to population growth and stagnant oil revenues). Despite the trend, Saudi nationals continue to have a standard of living far higher than citizens of non-oil-rich countries in the region. The provision of substantial social services, cheap housing, and subsidized energy and water costs also acts to prevent the creation of a true dispossessed class.

Does a mechanism exist for changing the static system of wealth distribution described above? Restrictions on access to wealth are primarily informal (but dominant), as substantial income accumulation and economic activity are at the discretion of the royal family. An informal but clear ceiling exists on income accumulation for non-

---

[10]James Adams, "Americans Face Rerun of Iran as Saudi Wobbles," *London Times,* July 28, 1996.

royals. The root cause of the situation is the political structure of the country. As long as the Sa'ud remain in power, the state of affairs is likely to persist.

The patterns of assumed loyalty to the royal family among the tribes and regions are a major determinant of access to wealth. Individuals from the Najd tend to be favored. Conversely, access is most difficult for the Shi'a because their loyalty is presumed suspect. While completely informal, these restrictions are ever present.

## Closure in the Social Realm

Status distinctions in Saudi Arabia are prominent and crucial to the standing and opportunities open to an individual in the kingdom. As befitting a monarchy, status is arranged in concentric circles, based on proximity and loyalty to the royal family. Not based on any clear ethnic criteria, status is a mixture of religious, tribal, regional, and economic factors. A status stratification map for Saudi Arabia might appear as in Table 6.3.

The royal family has absolute preeminence in the social hierarchy in the kingdom, followed immediately by the religious establishment. The royal family and the Muslim clerics co-exist in a symbiotic relationship, dating back to the close identification between the House of Sa'ud and the Wahhabi movement since the latter's beginnings in the 18th century. The core source of the royal family's high status is the simple fact that the House of Sa'ud conquered most of the

### Table 6.3

### Status Stratification

| Group | Status |
|---|---|
| Royal family | + + |
| Clerical class | + + |
| Najdi tribes | + |
| Other regions/tribes | − |
| Shi'a | − − |

+ + = high status; − − = low status (compiled on the basis of data presented earlier).

Arabian peninsula, established the kingdom, and now treats it *de facto* as if it were a personal possession. Because of the location of Islam's holiest shrines in the kingdom, religion is a central factor in its identity. The Sunni (Wahhabi sect) clerics are dependent on the royal family for their own high position in the kingdom and in the Muslim world. Other religious schools of Islam are not recognized in the kingdom in practice, and their clerics have no standing, if any legal existence at all. The clerical class is drawn strictly from key Najdi tribes, but especially from the Al al-Shaykh family, who have long represented the ideological/religious support to the royal family. In turn, the Wahhabi interpretation of the Islamic law is upheld by the kingdom's state apparatus (justice and court system).

Social status for every other group stems from its historically determined loyalty and identity with the House of Sa'ud. The various tribes in the Najd have been associated with the House of Sa'ud the longest and participated in the original alliance put together by Ibn Sa'ud that conquered most of Arabia. Traditionally, nearly all Najdis are followers of the Wahhabi sect of Islam.

Groups from other regions (such as Hijaz or Asir) rank below the Najdis, because they were associated with other sects of Sunni Islam and were conquered relatively recently; there are also longstanding rivalries between some of them and the House of Sa'ud. The Hijazis are not considered to be "pure Arabs" in the way the Najdis are. However, the Hijazis have considerable economic and commercial clout; Jeddah had been the commercial capital of the kingdom and has only recently been rivaled or surpassed by Najdi economic and commercial power and the rise of Riyadh.

The Shi'a of Saudi Arabia are unquestionably at the lowest end of the social order. Their religious sect is officially described by some Wahhabi clerics as apostasy (*rida* or *kufr*) and condemned as non-Islamic (*takfir*), which is (theoretically) even punishable by death. Shi'a exile groups claim they are not allowed to teach or study their faith, and have been prohibited from building Shi'ite mosques and community halls for nearly 70 years.

What are the implications of this status distribution? Status is an informal but powerful determinant of access to political and economic power. There is a whole range of subtle biases and assumptions about the proper social place of individuals, based on their group affinity. In the case of the Shi'a, the biases and assumptions amount to exclusion from many career paths. In any event, Shi'a access to advancement is limited to the private sphere (since there is little prospect of selection and/or advancement in the public sector), and even then it is primarily within their own region (the Eastern Province). Shi'a are excluded from at least 50 percent of Saudi universities and research institutes, and Shi'ite testimony is not permitted in court. Shi'ite shrines have been destroyed by Sunni activists in the past. Shi'a activists claim that even certain Shi'ite given names are prohibited.[11]

The other groups are also a target of the clerics, but not nearly to the same extent as the Shi'a. For example, the Hijazis are deprived of practicing officially their own brand of orthodox Islam (the Hanafi school) with their own clerics, but in practice the constraints are less harsh. Because of their low status and implicit concerns about their loyalty to the monarchy, the Shi'a tend to be seen as more loyal to Iran (the Shi'ite intellectual center) and treated as likely traitors by the rest of the Saudi society.[12] For example, after the Iranian revolution in 1979 and Saudi concerns that the Shi'a in the kingdom could be a fifth column, Shi'ites were systematically eased out of nearly all significant positions in the oil industry in the Eastern Province.

Is there a mechanism for change within the Saudi Arabian social stratification system? Status is ascriptive in that it is acquired primarily by birth and little movement is possible between status groups. The only exception is that connections to the royal family or certain privileged merchant groups can affect status. In this sense, some change of status is possible by way of succeeding at the professional level and/or showing loyalty to the regime. Nearly all restrictions are informal, in some ways making them more difficult to change (because they are hard to pinpoint). There is little interaction between groups, as one's social circle is narrow, discrete, and primarily limited to the extended family.

---

[11]Based on statements made to author Fuller by several Shi'a leaders in exile.

[12]Based on statements made to author Fuller by several Shi'a leaders in exile.

## Overall Assessment of Closure

Based on the information presented above, Table 6.4 summarizes the degree of closure (in an overall sense as well as in the political, economic, and social realms) experienced by Saudi Arabia's main ethnic groups. To reiterate, closure in Weberian terms refers to the "process of subordination whereby one group monopolizes advantages by closing off opportunities to another group." In the table, a group experiencing a "low" degree of closure has the most opportunities open to it. A group experiencing a "high" degree of closure has opportunities largely closed off.

The royal family is clearly privileged in all respects, to an extent seldom seen in the contemporary world. The clerics occupy a privileged position in the kingdom as the interpreters of God's law. The Najdi tribes (with some differences among them) are the main loyalists of the regime, and they show a pattern of privilege. The fact that all three groups face a low degree of closure does not mean that they are equal. The royal family is so far above the rest of the society that it does not really belong on a scale with them; it is included here only for reasons of completeness.

The non-Najdi tribes and regions differ in status among themselves, but in relation to the privileged Najdi groups, all of them experience a moderate degree of closure in the kingdom's social and political life. Among them, the Hijazis probably face the fewest obstacles at the economic level. The Shi'a face a consistent pattern of deprivation and experience a substantial degree of closure, especially in the

Table 6.4

Patterns of Closure by Group Affinity in Saudi Arabia

|                | Political | Economic | Social   | Overall  |
|----------------|-----------|----------|----------|----------|
| Royal family   | Low       | Low      | Low      | Low      |
| Religious elite| Low       | Low      | Low      | Low      |
| Najdis         | Low       | Low      | Low      | Low      |
| Hijazis        | Moderate  | Low      | Moderate | Moderate |
| Asiri Yemenis  | Moderate  | Moderate | Moderate | Moderate |
| Shi'ites       | High      | High     | High     | High     |

kingdom's political and social life. However, the kingdom's oil revenues have led to relatively generous social welfare policies, limiting the outward differences stemming from the various levels of privilege and deprivation.

The informal nature of the rules that uphold the social standing of groups in the kingdom, the absolute, personalized, and kinship-based system of governance, and norms of a traditionalist Islamic society all combine to make for a rigid stratification pattern. As long as the monarchy retains its current hold on power, there is little indication that the existing patterns of stratification will undergo any change.

A final ranking of groups along the lines of privileged to dominated—in relative terms—is in Table 6.5. The specific placement of ethnic groups on the privileged-dominated scale is not evenly spaced. The royal family is vastly above the other groups. The religious elite is also in a special position, *de facto* as a junior partner for the royal family in ruling the kingdom. The main regional/tribal divisions range from the privileged Najdis to the clearly dominated Shi'ites.

**Table 6.5**

**Ranking of Groups in Saudi Arabia**

| Privileged | Royal family |
|---|---|
| | Religious elite |
| | Najdis |
| | Hijazis |
| ↓ | Asiri Yemenis |
| Dominated | Shi'ites |

# TRANSFORMING POTENTIAL STRIFE INTO LIKELY STRIFE

As a prospective case study, this section focuses on the process of potential mobilization of one of the main groups against the state. The primary focus is on the Shi'a, and that mobilization process is sketched out in full. That is followed by a brief discussion of some other non-Shi'a-centered scenarios.

## PRIMARY SCENARIO:  SHI'ITE MOBILIZATION

The Shi'ites face the greatest level of discrimination and deprivation in Saudi Arabia on a group basis.  Out of all the Saudi substate groups, the Shi'a also come closest to being an "ethnic group" in the usual sense of the term:  the Shi'ites are a distinct group, having their own customs and inhabiting a compact area of the kingdom, with religious affiliation being the main "marker" differentiating them from the rest of the Saudi society.  The irony of the Shi'ite situation is that even though they inhabit the primary oil-producing region of Saudi Arabia (the largest petroleum producer in the world), the other regions of the kingdom see the effect of revenues from the oil indus- try.  In other words, the Shi'ites sit on top of the greatest oil reserves in the world, but not only do they see relatively few benefits from it, they are condemned by the Saudi clerical elite as "apostates" or "non-Muslims" and shunned and treated as likely traitors by the rest of the Saudi society.  The mixture contains the potential for a rebel- lion.  Indeed, during the past two decades, numerous acts of open dissent and violence (riots and terrorist bombings) by Shi'ite activists have taken place.  The scenario presented below sketches the poten- tial path of Shi'ite mobilization leading to a full-blown challenge to the monarchy.

## Incipient Changes

The combination of longer-term demographic and economics trends may prove troublesome for the kingdom.  Oil prices fluctuate sub- stantially from year to year, but trends in place point to low or mod- erate prices for oil.  New oil discoveries, the opening up of new areas to oil production, and advances in technology mean that, in the absence of OPEC regaining its 1970s-era effectiveness as a cartel, the world supply of oil remains sufficient to meet gradually increasing demand.  However, a low or moderate price of oil means a constraint on Saudi Arabia's economic growth (because of the country's re- liance on oil for revenues).  On the other hand, the natural increase of the population in Saudi Arabia remains high (estimated by the CIA at 3.42 percent in 1997) and, given the traditionalist Islamic outlooks, it is likely to continue at levels well above 3 percent.  The two trends are problematic in the long term, in the sense that they are bound to accentuate the competition for economic resources within Saudi

Arabia. Given the kingdom's social structure and the ever-growing size and appetites of the royal family, the equation boils down to a sharpening of competition among Saudi nationals for a share of a pie that is shrinking or at least increasing at a slower pace. In such a situation, the disadvantaged groups, most of all the Shi'a, will be increasingly squeezed out and face declining standards of living in an absolute and probably in a relative sense to other groups. Complicating the problem further, continuing investments in the petroleum-extraction industry mean economic activity in the primary Shi'a area, with its attendant pollution problems but without the material benefits accruing to the Shi'a. The trends put in place a source of concern among the Shi'a concerning the further deterioration of their social and political position in the country.

Environmental problems constitute another potential problem. The monarchy's favoritism toward the Najdis means that heavily subsidized agriculture (wheat, at many times the international market price) has flourished in the Arabian desert, rapidly depleting the water table in the Najd. Consequently, there may be pressure to divert water resources from the Eastern Province (the best farming region in the kingdom) to the Najd or even to push the Shi'a from their farms. Either of the two developments would sharpen the competition between the Shi'a and the Najdis.

In terms of sudden (or at least less-expected) changes, the primary economic shock might consist of a sudden and substantial drop in the price of oil (caused either by a worldwide recession or a technological breakthrough that would make the advanced economies less dependent on oil). Such an event would lead to a rapid drop in state subsidies and social services and would have a damaging effect throughout the Saudi society, but it would fall the hardest on the least privileged, namely the Shi'a.

In terms of a sudden political change, a royal succession is never an easy transition. Eventually, like any mortal, King Fahd will pass away from the scene. Even if the designated successor takes power without major problems, the initial period of the new ruler's reign is bound to be full of uncertainty just because of the personalized nature of royal rule. If the monarch is deposed as a result of an assassination (King Faisal was assassinated by one of his nephews in 1975) or a coup, the period of transition to a new ruler may be even more turbulent.  In any event, a leadership change in a monarchic system is always a crisis of the system.  Individuals with antipathy toward the system may see the crisis as an opportune time to pursue their own agendas.

A diminished level of support for the monarchy by the main loyalist forces (Najdis) would have the effect of suddenly changing the power balance in the kingdom.  Such a switch is not that far-fetched, and it could have origins in the fundamentalist Sunni opposition to the monarchy (evidenced by terrorist acts) combined with reduced loyalties among the Najdis in general because of rising expectations and the problematic economic performance.

The rise of a new Shi'ite religious leader abroad would amount to a major change.  The ascent of Ayatollah Khomeini after the Iranian revolution had an inspirational fallout on a small but significant number of Shi'a in Saudi Arabia.  With Khomeini's passing, no revolutionary Shi'a leader of comparable stature has emerged.  Should one arise in Iran, Iraq, or elsewhere, such a figure would provide a cohesion to the Shi'a that is currently lacking.

## Tipping Events

Galvanizing or tipping events might involve the arrest (or release, as several prominent Shi'ite clerics are in jail) of key Shi'ite religious leaders or any number of specific steps to clamp down on the public expression of Shi'ite religious practices; examples might be the closing of Shi'ite mosques or meeting places (*hussainiyyas*) or renewed denunciation of Shi'ism as heretical by the Sunni clerics.  The reinstitution of strict controls on Shi'ite travel abroad or in the kingdom carries with it a symbolic significance of singling out the Shi'a and may galvanize the group.

## Leadership

Leadership among the Shi'a is most likely to come from the clerics because of the moral authority of such figures.  Numerous Shi'ite clerics already in jail could serve as leaders, as could some of the clerics in exile.  Shi'ite leadership might emerge abroad because of the restrictions on all Shi'a in the kingdom, compared to the ability of the Shi'a abroad to write, publish, and disseminate information to their followers. Shi'ite centers in Iraq, Iran, Lebanon, and the United Kingdom (London) all have ties to and an impact upon the Saudi Shi'a. Religious leaders have the potential for a wider following, since they are seen as a source of political guidance. Moreover, these figures could draw on a built-in infrastructure of Shi'a meeting places. Shi'a religious practice has strong hierarchical elements, allowing for high-level leaders to have tremendous influence over their followers.

Secular Shi'a leadership would be more difficult because Shi'ite institutions are virtually nonexistent. Under less politically constrained conditions, a leading Shi'ite merchant family might be a source of secular leadership, but few such families are in evidence because of the pattern of discrimination.

## Resources and Organization

The Shi'a have limited financial resources available to them, especially when compared in a relative sense to the other main regional groups. The pattern of discrimination faced by the Shi'a in the kingdom has meant that few Shi'a have accumulated substantial wealth. Moreover, the isolation of the Shi'a from others in the kingdom means a lack of readily identifiable groups that may sympathize with or support a Shi'a mobilization (other than perhaps the case of opposition to the royal family among some fundamentalist circles in the Najd). The resource problems of the Shi'a are magnified by the absence of any clearly identifiable mechanism through which a leadership could "assess taxes" in the name of the group. Organizational skills are less of a problem than the fact that Shi'ite organizations are not permitted. Potential networks are currently weak, and the monarchy tries to keep them that way.

## Foreign Element

The foreign element is a strong factor in a potential Shi'a mobilization, and it may partially fill some of the shortcomings already identified above. The Shi'a of Saudi Arabia have religious ties with Shi'ite elements outside the country. Religious leaders have contact with senior clerics in Iran and Iraq. Some have ideological ties and contact with the Shi'ite Hizballah organization in Lebanon. Family ties link many Shi'a in Saudi Arabia with the Shi'a in Bahrain (the latter have engaged in protest and insurrection against the Bahraini regime). If the Shi'a were to come to power in Bahrain, even a moderate government could not disavow all interest in the fate of the Shi'a in Saudi Arabia. The same applies to the possibility of the accession to power of the Shi'ite majority in Iraq. Clearly, there is a multitude of ties between the Shi'ites of Saudi Arabia and Shi'ite organizations that have expertise in conspiratorial and terrorist activities. And there is a thin line between sympathy along the co-religionist/co-ethnic lines and the use of Shi'ite organizations for their own ends by the neighboring countries, with Iran being the best case of this.

Saudi Arabia's land and sea borders with Yemen, Oman, the United Arab Emirates, Qatar, and Kuwait remain in dispute (though most of them are in the process of being negotiated or demarcated). The disputes are minor, but during the 1990s they did lead to armed skirmishes between Saudi forces and Yemeni and Qatari guards. Iran, Iraq, Syria, Jordan, Egypt, or Yemen all have the power and geopolitical interests to assist potential internal opposition groups within Saudi Arabia. Iraqi forces invaded Saudi territory during the Gulf War, and Iraq could generate a border dispute if it chose to do so. Iranian support for terrorist organizations in Saudi Arabia is widely acknowledged. In principle, Iran could decide to provide direct support to the Saudi Shi'a. So could elements of the Shi'ite community in Iraq, especially if it were to come to power. Hizballah in Lebanon also constitutes an important potential conduit of armed aid to the Saudi Shi'a. A specific spark that may lead to the accentuation of the role of the foreign element in a potential Shi'a mobilization in Saudi Arabia would be heightened U.S./Saudi tensions with Iran.

## Overall Assessment of Mobilization

Given the shortcomings in the elements needed for Shi'a mobilization and the enormous resources (in both a relative and an absolute sense) that the monarchy can marshal against them, a crucial aspect of the effectiveness of any Shi'a mobilization is a weakening of the Saudi monarchy. This can take the form of a succession crisis or a reduced level of support (or more pronounced dissatisfaction) with the monarchy among its main loyalists and supporters. A potential Shi'a mobilization is lacking in resources and organization, but these problems might be offset through a greater involvement of the foreign element. Effective leadership would play a crucial role in the mobilization process, as it would limit the resource problems and attract foreign support.

## OTHER SCENARIOS

The Hijazis and perhaps even the Asari Yemenis may be worth examining in a manner similar to the Shi'a assessment. While potentially dangerous to the monarchy, neither of these two scenarios falls in the category of ethnically based mobilization as defined earlier. Instead, they amount to potential mobilizations on group bases but more along the lines of regional differentiation in a highly socially stratified country where one region is distinctly favored by an absolutist regime. Since the issue at hand is potential ethnic mobilization, the only group that comes close to fitting that description in Saudi Arabia is the Shi'a.

Nonetheless, it is important to note that rigid stratification in modernizing societies (and Saudi Arabia is an uneasy mix of the most modern and the most traditional) tends to be problematic. Eventually, if a system of privilege and discrimination along regional lines is kept in place by state policy, the initially slight regional variations may lead to the emergence of ideologies that expressly claim "ethnic" differences (and use historical examples found in every society to "prove" the point) and/or aspire to differentiated (for example, autonomous) status because of the internalization of the prejudices upheld by the state. Such a process is generational, but it tends to quicken in times of growing economic grievances toward the ruling group. The tribal and regional allegiances that remain strong in Saudi Arabia may or may not take such a path, but signs pointing in that direction are worth monitoring.

Finally, the case of Najdi-led opposition to the monarchy, either alone or combined with Sunni fundamentalism, is a distinct threat to the regime. Such a scenario does not merit an examination from the perspective of an ethnically based mobilization, but it could be analyzed in a systematic fashion, perhaps using some of the elements and categories from the framework used here.

## ASSESSING THE STATE

### Accommodative Capability

How inclusive and responsive are Saudi Arabian political structures to the popular will? The political institutions in Saudi Arabia are expressly exclusive by design, and the Saudi leadership makes no apologies about it. There are no means to hold political authorities accountable to popular will. The political system is monarchic and, as such, there are no voting rights or elections. There is no constitution, and the Islamic law serves as the basis of the official justice system in the kingdom. There are few constraints upon the king, and all participation in affairs of the kingdom is at the pleasure of the royal family. The level of exclusion and inclusion varies according to the group, with the Najdis most favored and the Shi'a most excluded. Almost all exclusion is informal and based on group-level assumptions about loyalty to the monarchy.

The few constraints upon the royal family, and any hint of accountability, come from two sources. One, the recently set up Consultative Council, appointed by the king, in principle has the right to ask ministers to come and justify their policies. In practice the council is timid and does not question policy in any depth or hold the policy executors accountable except at the most politically safe technical level. Two, the moral authority of the clerics limits the regime. However, the senior clerics have become largely subordinated to the political order and openly state that it is against Islamic law to oppose the state (some radical clerics challenge such an interpretation of the Islamic law). The king appoints the members of the highest Islamic authority and can make sure that the top clerics play a role in upholding the monarchy.

Is there potential for change to be found in the Saudi Arabian political structure of late 1997? The political structure is absolutist and has no formal mechanisms for conflict resolution and institutional change. The king's decision, usually justified on the basis of the Islamic law, is the final word.

In one semblance of a conflict-resolution mechanism, the royal family holds open audiences whereby, in principle, citizens may bring their grievances to the highest level for consideration. The practice allows the royal family to hear a narrow segment of the *vox populi,* but the decisions of the emirs at such meetings are arbitrary (even if well intentioned) and are usually not professionally informed or debated. Theoretically, the issues brought up at such meetings could touch upon aspects of the political system.

The only institution that has the seeds of potential change of the political structures is the Consultative Council, though it must evolve a long way before it will have any impact. Conceivably, over time the council could take on greater powers—perhaps a technical challenge to some nonroyal minister or a more active proposal of legislation to the cabinet—but such a change could only occur with the consent of the king. In short, the system is one of the most inflexible, unresponsive, and exclusive in the contemporary world.

In terms of prevailing norms of governance, the monarchy is not tolerant of opposition. The co-opted clerics and the justification of the Islamic law provide the mechanisms to squash all dissent. Nonetheless, the regime does not rule with an iron fist in the same fashion as the Syrian or Iraqi regimes. The monarchy would rather co-opt than punish, marginalize and gratify (through a policy of buying off enemies by making "offers that cannot be refused") rather than dictate. With large carrots as well as sticks, only the uncommon ideologue prefers to face the stick. When the carrot fails, however, the stick (or the sword) can be harsh.

The monarchy's assumptions about individuals' loyalty on the basis of group origins shows in its favoritism of the Najdis and the marginalization of the Shi'a. The outlook indicates thinking in terms of group-defined collectivist norms.

What is the level of cohesion among the ruling elites? Only anecdotal evidence and inferences on the basis of past events shed some light on the inner workings of the royal family. Previous experience shows that informal consultation within the royal family is ongoing and ever present. Such consultation has led to the deposing of a king by other members of the family (King Sa'ud). As in any closely knit kinship-based system of rule, court intrigues are the order of the day.

The royal family shares a broad consensus about the policies of the government and the nature of the society, though anecdotal evidence indicates some fears about the family's ability to maintain itself in power in the long term.  Known policy differences include the degree of desirable closeness to the United States and the preferred course of action on domestic issues such as corruption.  There may be some disagreement at lower levels of the family relating to the share in the benefits for members of the ruling family and possible concerns over the bureaucratic rigidities of the system.  But although the royal family may appear divided when examined from within, when dealing with outside threats the group is likely to be highly cohesive.

In conclusion, the institutional structure of the Saudi monarchy is exclusive, inflexible, and has little tolerance in governance.  The kingdom is not responsive to internal pressures and has little potential for change.  Cohesion among the ruling elite in support of the system is high.

## Fiscal and Economic Capability

The Saudi monarchy presides over a massive income from oil revenues, making it a classic rentier state, i.e., living on "rents" in which the state's major function is the distribution of largesse over which it has almost total monopoly.  Saudi oil reserves, the largest in the Middle East, will extend revenues through the next century, even at the current rate of production.  Saudi Arabia's proven recoverable petroleum reserves were sufficient to maintain production at 1995 levels for over 80 years (equivalent to about one-quarter of the world's proven oil reserves).  Because of high state spending levels established in the 1970s (at a time when oil prices were high), the state has had problems in cutting its expenses to match the drop in revenues derived from oil exports (oil income dropped by nearly one-half in the 1980s).  As a result, the state has been forced to cut back on services, benefits, and subsidies, and it has even defaulted on payments to some foreign contractors.

The macroeconomic indicators are stable even though the economy has registered only small growth or slight declines annually for over a decade.  In the five-year period 1992–1997, Saudi Arabia's overall balance of payments has been mostly positive, even though the current account balance has been largely in the red. The kingdom has followed a pattern of light deficit spending in the 1990s (less than 1 percent of GDP). Detailed information on state debt is not available. Saudi Arabia's gold reserves have fluctuated but remain strong. The kingdom also has substantial loans it could call on from international lending institutions. See Table 6.6.

The state has no mechanism in place to extract resources from the society, but society is not a source of income given the nature of the rentier state. There is no taxation in Saudi Arabia and none is envisioned in the foreseeable future. Its resource base stems from revenues derived from oil exports. Because the kingdom's economic performance and availability of resources rely on and mirror the world demand for oil, a worldwide recession would present problems. However, demand for oil is only partially elastic and, as the world's largest oil producer, Saudi Arabia is guaranteed a steady stream of revenues.

There is potential for the reappropriation of funds within the state budget. Public administration and some social services have declined slightly, but defense and security expenditures have remained at approximately 33 percent of the state budget during the past five years. There seems to be enough leeway for substantial shifting of funds on a short-term basis. See Table 6.7.

The royal family and its supporting elite are extraordinarily wealthy (among the wealthiest in the world). Informal provision of state-salaried jobs and distribution mechanisms within the key tribes provide the chief loyalists with substantial income.

### Table 6.6

### Saudi Arabian International Reserves
### (in US$ million in December each year)

|                             | 1994  | 1995  | 1996  |
|-----------------------------|-------|-------|-------|
| Gold                        | 235   | 239   | 231   |
| IMF special drawing rights  | 607   | 666   | 692   |
| Reserve position in IMF     | 882   | 854   | 807   |
| Foreign exchange            | 5,888 | 7,101 | 5,295 |
| Total                       | 7,613 | 8,861 | 7,025 |

SOURCE: IMF, International Financial Statistics.

**Table 6.7**

**Saudi Arabian Government Expenditures (in 000,000 riyals)**

|                                             | 1994    | 1995    | 1996    |
|---------------------------------------------|---------|---------|---------|
| Total                                       | 160,000 | 150,000 | 150,000 |
| Defense and security                        | 53,549  | 49,501  | 50,025  |
| Public administration and other government spending | 40,530  | 39,706  | 37,952  |
| Services and subsidies[a]                   | 65,398  | 60,317  | 61,608  |

[a]"Services and subsidies" include the following categories: human resource development, transport and communications, economic resource development, health and social development, infrastructure development, municipal services, local subsidies.

SOURCE: IMF, *International Financial Statistics.*

In sum, in an absolute sense, the fiscal capacity of the Saudi monarchy is strong. The monarchy has access to enormous resources, in both an absolute and especially a relative sense (to a group such as the Shi'a). Despite Saudi Arabia's economic difficulties during the past decade, it would be absurd to describe the state as facing a financial crisis or lacking financial resources to meet its own political priorities. The royal family has total control over the budget and is capable of making substantial shifts in resource allocation at will, if the welfare of the family depends on it. A key political question is how much of its own income the royal family might be willing to sacrifice in order to make more funds available to other elements of society in the interests of buying social peace and tranquillity.

## Coercive Capability

The royal family maintains total control over all apparati of violence within the state, and is subject to virtually no institutional restrictions on its use. The king is the head of state, the prime minister, and the commander-in-chief of the armed forces. The king's designated successor and the first deputy prime minister, the Crown Prince, is the commander of the National Guard. The second deputy prime minister (another prince) is the defense minister. Another prince is the minister of the interior (police and domestic security forces). In principle, the senior clerics have a moral voice in policies relating to the application of state violence, but most of the clergy has been co-opted to the royal family's needs.

Who serves in the apparati of violence? The regular armed forces are composed of well-paid volunteers, usually of urban backgrounds. Information about the armed forces, its officer corps, and particularly its attitudes toward political developments is highly sensitive, and there is only fragmentary and anecdotal evidence on which to base assessments. The state security bodies, the intelligence services, and the National Guard are manned almost exclusively by Najdi loyalists (drawn from loyal tribes from the Najdi heartland, all carefully balanced off against one another). There would be very little sympathy on their part toward any non-Najdis, particularly the Shi'a, who might challenge the monarchy's authority in the kingdom.

What norms exist with respect to the use of violence domestically? The Saudi monarchy has never wavered in the application of force to quell public disorder. That said, the general wealth of the regime has allowed it to buy off most opponents, and the monarchy has not had the need to practice violence against the population on a broad basis.

Compared to the libertarian foundations of U.S. law, Islamic law in its strict interpretation (as applied in Saudi Arabia) is harsh and unforgiving. Saudi Arabia tends to lead the world in the application of the death penalty, and there is only limited experience with ideas of human rights, as enshrined by the UN.

Are the Saudi apparati of violence suitable for domestic use? The forces most suitable for domestic use include the Public Security Force, Ministry of the Interior Forces (intelligence and police forces), and the National Guard. All of them have the training and equipment to maintain public order. The National Guard, the preeminent internal security force and counterweight to the regular military, is well funded and equipped.[13] Its strength seems to match that of the military, and its training has been provided by, among others, the

---

[13]The phenomenon of having in effect two armed forces (the National Guard and the regular army) as a check on each other forms a main component of what one analyst has dubbed "coup-proofing" of the regime. James T. Quinlivan, "Coup-proofing: Its Practice and Consequences in the Middle East," *International Security*, Vol. 24, No. 2, Fall 1999, pp. 131–165.

U.S. Vinnel Corporation. The regular armed forces are not well suited for domestic use, since they contain substantial heavy conventional forces. But they could be used to back up the other domestic security formations in operations against widespread disturbances.

In conclusion, the Saudi monarchy has absolute control over an extensive coercive apparatus. There are no formal constraints on the use of force and only the vaguest of informal constraints. The apparati of violence appear to be well trained and armed and are politically trustworthy.

## STRATEGIC BARGAINING

Based on the framework presented in Chapter Two, this analysis of the potential mobilization of the Shi'a and the capabilities of the Saudi monarchy to deal with such a mobilization provides a way to think about the range of possible interactions between them. As in the previous three case studies, the group and state types are categorized on the basis of their capacities, and the matrices assist in conceptualizing their likely patterns of interaction.

### Measuring the Group's Capacities

Concerning the leadership of the Shi'a, the profile of Shi'ite clerics fits all the conditions of a strong leadership. They are certain of the rightness of their cause and have a clear vision of what they wish to accomplish. Their interpretation of Islam is supported by a major player in the Islamic world (Iran). The leadership has not hesitated to take risks and has paid the price for it. A clear leader emerging from the clerics would have high standing in the group. Thus, the assessment of leadership that a potential mobilized Shi'a group is likely to have is "strong."

With respect to resources, internal support for the Shi'a in Saudi Arabia is weak, especially in comparison to the enormous wealth of the monarchy. The monarchy has been successful in marginalizing the Shi'a economically and limiting their ability to organize networks within Saudi Arabia. In addition, the Shi'a cannot count on support from other groups in Saudi Arabia. There is an extensive potential support network from abroad, but its suitability is not all that clear and is still limited in relation to the resources available to the monarchy. Thus, the assessment of resource support that a potential mobilized Shi'a group is likely to have is "weak."

Regarding popular support for the Shi'ite mobilization, this capacity is not as clear-cut as the others for definitional reasons. A potential Shi'ite mobilization would build on an already strong group cohesion (the survival of the Shi'a as a community in a hostile environment is testimony to its existence). However, the Shi'a lack sympathy or potential for support from other groups in the kingdom. This fact strengthens the in-group feeling of the Shi'ites but is a weakness in any localized Shi'a mobilization against the Saudi monarchy. Even if the mobilization took place in a situation of a more general decrease of support for the monarchy in the kingdom (as stipulated in the incipient changes), the move could result easily in greater cohesion between the monarchy and the other groups against the Shi'a rather than support from the other groups for the Shi'a. Popular support from co-ethnics and co-religionists may provide greater resources and organizational skills, but it does not change the basic problem that the Shi'a are isolated in Saudi Arabia. Thus, the assessment of popular support that a potential mobilized Shi'a is likely to have is "weak."

Based on these assessments, the potential mobilized Shi'ites are judged to be a type G group. The capacities of such a group are as follows:

Accommodative: low;

Sustainment: low;

Cohesiveness: low.

Note that if popular support is assessed as broad, then the group changes to type C, whose sustainment capacity remains low but whose accommodative and cohesiveness capacities change to high.

## Measuring the State's Capacities

Concerning the leadership of the state, the royal family has grown used to its position and has a clear goal to maintain power and perpetuate the status quo. Its absolute hold on power makes any questions about losing popularity moot. Its strategy is also clear: as long as it keeps the clerics a part of the ruling structure and retains a core of loyalists, it will continue to profit from the oil revenues. Thus, the assessment of leadership is "strong."

The fiscal position of the Saudi monarchy is strong. The reduced profits from oil exports during the last decade notwithstanding, Saudi Arabia remains in an enviable situation of having virtually guaranteed substantial revenues for the foreseeable future. Because of Saudi Arabia's importance to the world economy, the monarchy can also expect generous loans if they became necessary. Thus, the assessment of fiscal position is "strong."

As an absolute monarchy, Saudi Arabia is one of the most exclusive regimes in the world. Political rights, as commonly understood in the United States, do not exist in Saudi Arabia. There are no formal and only the vaguest of informal checks upon the regime. The regime is intolerant of dissent and its system of law is strict, harsh, and unforgiving. Thus, the regime is assessed as "exclusive."

Based on these assessments, the Saudi monarchy is judged to be a type D state. The capacities of such a state are as follows:

Accommodative: low;

Sustainment: low;

Coercive: high.

## Outcome of Bargaining and Preferences for Violence

Based on the matrix showing the preferences of the mobilized group, a type G group has the following preferences toward a type D state: (1) intimidate, (2) negotiate, and (3) exploit.

And based on the matrix showing the preferences of the state toward a mobilized group, a type D state has the following preferences toward a type G group: (1) repress, (2) exploit, and (3) negotiate.

Note that if the group type is C, due to an assessment of popular support as broad, the group's preferences remain the same: intimidate, negotiate, exploit. The state's preferences also remain the same: repress, exploit, negotiate.

Comparing group and state preferences leads to the conclusion that the potential for violence in the dyadic encounter between the mobilized Shi'a and the Saudi Arabian state is high. The primary strategy for both sides is to turn to violence.

The state has a preference for a strategy of repressing the challenger, with a hedging strategy of exploiting the weaknesses within the mobilized group. In practical terms, the latter may mean a forceful crackdown and intimidation of the Shi'ite activists while using economic incentives to reduce in-group support for the activists. But the primary strategy is to deal with a Shi'a mobilization by force. The strategies do not change even if the Shi'ite mobilization were to obtain greater support within Saudi Arabia (leading to coding the group as type C).

The type of state that Saudi Arabia fits (type D) is the most violence-prone of all eight state types. In dealing with the variety of challenging groups, a type D state has a primary strategy of using violence six out of eight times. The results are not accidental but based on logical connections between resources and strategies. In practical terms, the regime is not constrained in any significant fashion from using force, it has the wealth to build up a powerful apparatus of violence, and it is determined to stay in power.

The preferred strategy of a mobilized Shi'a would be to intimidate the regime, with a hedging strategy of negotiation. In practical terms, the preference for violence may mean a sudden armed uprising, made even more likely as a strategy because of the expectation of foreign support to deal with an expected Saudi crackdown. But the juxtaposition of violent and peaceful paths as the preferred and hedging strategies is interesting and in contrast to the state strategy of force or threatened force in both the preferred and hedging strategies. The contrast illustrates the power difference between the group and the state. The nature of the regime and its preponderance of power leave little recourse for the group except to take up arms. But the group would be alert for any signs that the regime might be willing to negotiate (for it would like to see a negotiated outcome but does not believe it to be a realistic option under the circumstances).

Quite aside from the above, the type of group that a mobilized Shi'a fits (type G) is the most violence-prone of all eight group types. In dealing with a variety of state types, a type G group has a primary strategy of turning to violence five out of eight times. Put in terms of the expected Shi'a mobilized group characteristics, the group has little going for it other than a devoted, bordering on the fanatical, leadership. The group lacks the resources to take on the state, and the resonance of its potential mobilization is limited. The above does not change even if the group is assessed to have wider support (from other groups in Saudi Arabia) and categorized as group type C, since the state type is so inflexible and resistant to consider choices other than violence.

## Revisiting the Evaluation

There is much to be said for the portrayal of a course of events between a potential mobilized Shi'a and the Saudi monarchy as expressed in the terms above. Based on the evaluations earlier in this chapter, there is little question that a mobilized Shi'a would be treated inherently as an enemy of the Saudi monarchy and would feel the full brunt of the Saudi internal security forces. Conversely, there is little question that the Shi'a have few alternatives to violence, even though they would prefer to pursue their group aims by peaceful means.

Since the Iranian revolution and the Shi'ite riots in the Eastern Province, activism on behalf of the Saudi Shi'a has been carried on largely abroad, especially by the Reform Movement. The monarchy has harassed some Shi'ite clerics and jailed others (and then it has negotiated with the Reform Movement about the release of the jailed clerics in return for the movement ceasing its "dissident" activities). Individual Saudi Shi'ites have turned up fighting in a variety of Muslim "holy wars," and a number of Shi'ite terrorist attacks in Saudi Arabia have taken place. All of the above shows a substantial level of discontent among the Saudi Shi'a, but their mass mobilization has not taken place. Perhaps the Shi'a intuitively realize the costs of such an action, for according to the analysis conducted here, they would pay a hefty price for an open and mass-based challenge to the monarchy.

The state has shown no tolerance for any mass-based challenge, and it has worked assiduously and effectively to prevent its occurrence through co-opting, buying off, or isolating the malcontents.  The regime's dealings with the Reform Movement illustrate a pattern of clever bargaining with a mixture of threats and incentives, all from a position of strength.  Indeed, the regime has monitored the Shi'a closely and tried to encourage splits within the Shi'ite leadership.  The regime has had the means and the skill to head off any mass-based opposition.  In this sense, a challenge by a mobilized Shi'a against the monarchy has all the elements of being violent and being met with violence (as shown in the analysis) but the Saudi regime has been successful so far in preventing such a challenge from materializing.  It may continue to do so for a considerable period.

For the purposes of tracking the evolution of intergroup relations in Saudi Arabia, and especially the relationship between the Shi'a and the Saudi monarchy, the following trends need special monitoring by intelligence analysts:

- The ability of the Saudi regime to continue to control dissent is dependent on its oil revenues.  A change in Saudi Arabia's position as the major oil supplier in the world (due to discovery of large recoverable reserves elsewhere or technological advances that diminish the importance of oil), a cut in its share of the market (due to increased production by other suppliers or disasters in the Saudi oil fields), or a major drop in the price of oil (especially if caused by a worldwide recession) all would have potentially far-reaching impact on the regime's domestic stability.  In terms of the model, serious and sudden economic difficulties for Saudi Arabia would imply a move toward state type G, another violence-prone type.  That in itself would not imply any change in preferred strategies toward the mobilized Shi'a, but it might create other domestic difficulties for the monarchy that the Shi'a could use to its own advantage.

- The monarchy's ability to manage internal tensions as a result of rising expectations and the regime's decreasing ability to satisfy them are already a problem, and it is likely to grow in importance in the long term.  Moreover, dissatisfaction is likely to be more difficult to control by the regime because of technological advances (the Internet), access to foreign travel and media, and the presence of substantial numbers of Western-educated personnel.  A weakening of support among the regime's loyalists may provide an opportunity for more assertive actions by the Shi'a.

- A succession to the throne would be a crisis for the monarchy.  Although a designated successor exists, there is no guarantee that the removal of the king from power, whether through assassination or natural causes, will not spark in-fighting within the royal family over the right to be the king.  In terms of the model, uncertainty at the top of the Saudi hierarchy means potentially a movement toward state type H, a more peacefully inclined type.  Depending on the circumstances surrounding a succession, the Shi'a (and other groups in the society) may become more assertive.

- The evolution of the Consultative Council into a body that wields even a measure of real power or control over the Council of Ministers would be a watershed in Saudi political development.  Such an evolution borders on the improbable at this stage, but specific events could propel it forward.  It would mean a move toward a more representative and inclusive state and, as such, would imply a different role for the monarchy (and the clerics) in the kingdom.  In terms of the model, this would represent a move toward state type A, a less violence-prone type than the one currently in place.

- The emergence of a clear leader among the Shi'a clerics with strong appeal to Shi'ites in Saudi Arabia and elsewhere in the Persian Gulf may have a dramatic impact on the Saudi Shi'a, leading to their mobilization.  The result of a confrontation with the Saudi regime is likely to be violent and may have wider repercussions in the Gulf.

Thinking about the dynamics of intergroup relations and potential for unrest in Saudi Arabia, the framework presented above makes apparent the following intelligence needs and information gaps:

- A better demographic profile on the Saudi Shi'a, including their organization and wealth;

- More information about the Saudi Shi'ite leadership, including a profile of key figures and their views on the future of the Shi'a community in Saudi Arabia;

- More detailed information about the activities of the Saudi Shi'ite dissident organizations in exile, including their contacts with the Shi'ite community in Saudi Arabia and their influence upon them;

- Tracking the rise of charismatic figures among the Shi'ite clerics and their standing among the Shi'a in Saudi Arabia;

- Monitoring the relations between the Shi'a and other groups in Saudi Arabia, including contacts between the Shi'ite and non-Shi'ite political opposition;

- Extent of foreign ties and support to the Saudi Shi'a, especially from Iraq and Lebanon and the receptiveness of Saudi Shi'ites to foreign assistance;

- Information about disagreements and attitudes within the ruling family on domestic stability and accommodation with the political opposition;

- The extent of Saudi intelligence in penetrating the Shi'a community, primarily as an indication of the ability of the Saudi regime not to be surprised by the Shi'a;

- More information on the Saudi National Guard and the armed forces, including the attitudes and composition of the officer corps, and their training and readiness for domestic security.

Close attention to these issues may enable early identification of any impending turn toward internal strife in Saudi Arabia.

## FINAL OBSERVATIONS

The framework presented here examines a potential conflict situation in Saudi Arabia through a dyadic comparison of the Saudi Shi'a and the Saudi monarchy. The framework provides a tool to think abstractly about the mobilization patterns and challenges to the monarchy. It is meant to provide insight into the strategic preferences available to the main actors under certain "what-if" conditions. As such, the model aims to be parsimonious, and its narrowing of the crucial questions to the main dyad illuminates the logically derived preferences of both actors.

The results show that if the conflict were to take place within the bounds stipulated, it would almost certainly be violent, as both sides would see the use of violence as a preferred strategy. Indeed, the matchup is noteworthy in that it places the most violence-prone state type against the most violence-prone group type. The situation is hypothetical, and its elaboration in some detail in no way implies that such a conflict is likely to materialize. There is nothing inevitable about the conflict coming about, and many indicators point to the likelihood that the Saudi monarchy can prevent the mobilization of the Shi'a for the foreseeable future.

The model is explicitly parsimonious and does not capture the various intergroup relations that may be important in the emergence of intrastate conflict in Saudi Arabia. For example, greater assertiveness along tribal and/or regional lines may play a part in weakening the regime. However, given the largely isolated status of the Shi'a in Saudi Arabia, the issue may not be crucial.

Since the model was applied to a prospective case, its accuracy in anticipating the likely strategies that the challenging groups and the state are likely to follow is, as yet, unverifiable in an empirical manner. Just as in the Ethiopian case study, the purpose of applying the model to Saudi Arabia was to provide a series of hypothetical but likely situations and determine the evolutionary paths for both the state and the group that would result in the most violence-prone confrontations. Thus the model's usefulness lies in illustrating the "dangerous dyads."

As with all the case studies presented here, the model is a tool for structuring analysis. As such, it provides a shortcut to thinking—in conceptual terms—about the likely paths to violence. But the Saudi Arabian case shows clearly that difficult judgments and assessments remain for an analyst to make. For example, the difficulty of assessing the availability of resources and the level of popular support for the Shi'a may change the outcome of the analysis. The matrices that are a part of the model provide logically deduced options open to the group and the state, but the preferences cannot be applied in a mechanistic fashion.

## AFTERWORD

In an overall sense, little has changed in 1998–1999 to alter the assessments contained in this chapter. Two recent trends, hinted at in the analysis, have become more pronounced.[14] Politically, King Fahd has become less involved in decisionmaking due to ill health. Crown Prince Abdullah has assumed more power, though he relies heavily on consensus within the ruling family. Abdullah is more conservative in temperament and behavior than Fahd, which increases his popularity with many ultraconservative Saudis, particularly the Najdis. This shift, however, is not likely to change greatly the status of Saudi Shi'a or regional groups.

Economic problems also have diminished. As the price of oil bottomed out and regained its earlier levels, the kingdom's ability to dispense largesse has recovered. In more general terms, state capacity has increased. Long-term problems still loom, owing to the steadily increasing size of the royal family and its members' continued expectations for the "good life." But for the near term, the royal family is in a better situation than it was in the mid-1990s.

---

[14]For another recent discussion of the prospects of political violence in the Persian Gulf region (including Saudi Arabia), see Daniel L. Byman and Jerrold D. Green, *Political Violence and Stability in the States of the Northern Persian Gulf,* Santa Monica, CA: RAND, MR-1021-OSD, 1999.

## ANNEX: DEMOGRAPHIC CHARACTERISTICS OF
## SAUDI ARABIA IN 1997–1998

The following information about population characteristics is based upon the situation in Saudi Arabia at the beginning of 1998. The information presented here is the basic reference for the analysis in this chapter.

**Name:** Kingdom of Saudi Arabia.

**Capital:** Riyadh.

**Nature of government:** Absolute monarchy, with no legislature or political parties. An advisory body, the Consultative Council, with members appointed by the king, was established by royal decree and inaugurated in December 1993. The king rules in accordance with the Shari'a (Islamic law). By custom, the king is also the prime minister and, in that capacity, appoints and leads the Council of Ministers. The Council of Ministers acts as an instrument of royal authority in legislative and executive matters. The council makes decisions on the basis of a majority vote, but to be implemented, a decision requires royal sanction.

**Organization of the state:** A system of provincial government was set up in 1993–1994. The country has 13 provinces (mintaqah): Al Bahah, Al Hudud Ash Shamaliyah, Al Jawf, Al Madinah, Al Qasim, Ar Riyad, Ash Sharqiyah (Eastern Province), Asir, Hail, Jizan, Makkah, Najran, and Tabuk. The provinces are divided further into a total of 103 governorates. Local government organs consist of the General Municipal Councils (in several towns) and tribal and village councils (led by sheikhs).

**Date of constitution:** None. Rule is according to Islamic law (Shari'a). King rules by issuing decrees.

**Population:**  20.1 million (CIA estimate, July 1997).  The figure includes 5.2 million foreign nationals residing in Saudi Arabia.  Latest census figures (September 1992) gave a total of 16.9 million (of which 4.6 million were foreign nationals).  Estimated population growth rate is 3.42 percent (CIA estimate, 1997).

**Major ethnic groups:**  All Saudi nationals are "Arab."  Approximately one million people are Yemenis, with varying degrees of legal status (those in Asir province are subjects of the kingdom).  Tribal loyalties remain strong, although they have lessened as a consequence of settlement and urbanization of the population.  Kinship ties are also significant.  The principal tribes are the Anayzah, Bani Khalid, Harb, Al Murrah, Mutayr, Qahtan, Shammar, and Utaiba—all in Central Arabia.  There are at least fifteen minor tribes as well, including the more weakly defined tribal groupings outside the central Najd area.  Within the central heartland of the Najd, tribal rivalries persist; tribes such as the Utaiba and Rashid still harbor latent grudges against the ruling family.  Hierarchies among tribes exist and they are determined, in part, by their closeness to the royal family.  Statistics on the tribal makeup of the country are exceedingly difficult to find and are discouraged by the regime since they are politically sensitive.

**Languages:**  Major languages: Arabic.  There are slight differences in dialect that distinguish the Najd heartland of the kingdom from the Eastern Province, the Hijaz, and the Asir.

**Religions:**  Islam.  About 85 percent of the population are Sunni Muslims, and most of them belong to the strictly orthodox Wahhabi sect.  The Wahhabi movement dates back to the 18th century and originates from the Najd region of Arabia.  The conquest of Arabia by the Najd-based House of Saud in the 1920s led to the imposition of Wahhabism as the official faith of the country.  While the main schools of Sunni Islam are officially recognized in Islamic law, any non-Wahhabi practice in Saudi Arabia is viewed *de facto* improper.  Non-Wahhabi clerics have no official standing and have almost ceased to exist in the kingdom.  The Shi'a represent a distinct minority whose religion in Wahhabi eyes is not considered legitimate Islam.  Their numbers are much disputed, though they number at least 200,000 to 400,000.[15]  Shi'a activists claim that half a million Shi'a hide their status and pretend they are Sunni because of the persecution and disadvantages associated with Shi'ism.  In their main region of residence (the Eastern Province), the Shi'a represent approximately 33 percent of the population.  Altogether, Shi'a activists claim one million people, or even 25 percent of the population,[16] though the more generally accepted figure seems to be about 15 percent.  The latter figure would seem to support the claims by Shi'a activists of the existence of substantial numbers of "hidden Shi'a."  Foreign nationals residing in the kingdom include adherents to a variety of religions, including Christianity.

---

[15]*Saudi Arabia Handbook,* Federal Research Division, Library of Congress, Washington, Internet Edition.

[16]*Mideast Mirror,* August 27, 1996, p. 15.

# Chapter Seven

# FINAL OBSERVATIONS

## Thomas S. Szayna, Ashley J. Tellis, and Daniel L. Byman

The theoretical framework presented in this report provides a systematic understanding of the phenomenology of ethnic mobilization and its potential for violence. The specific utility of the model is to structure the analysis, delineating and explaining the specific rationale for anticipating violence. As the case studies show, the model structures the analyst's work but does not eliminate the need for her informed but subjective judgments throughout the process.[1]

## WHAT DO THE CASE STUDIES SHOW?

The retrospective case studies show that the model can help analysts predict the likelihood of conflict with a reasonable degree of confidence. The model also helps analysts place seemingly separate events, such as a foreign crisis, a decline in regime reserves, or the arrest of a charismatic figure, into a unified whole. As the prospective case studies show, using the model allows analysts to understand better today how a wide array of possibilities will affect intercommunal relations tomorrow. Through the model, analysts can ask "what-if" questions by varying key factors and, in so doing, determine which events might lead to conflict and which might lead to

---

[1]As an analyst noted recently, assessing threats requires "laborious, methodological, rigorous analytical work . . . [and] with the exception of a few notable pockets of excellence, the [U.S. intelligence] community would appear to need a boost in all three departments." Mary O. McCarthy, "The Mission to Warn: Disaster Looms," *Defense Intelligence Journal*, Vol. 7, No. 2, Fall 1998, p. 20. If that assessment is true, then the model presented here may assist with that boost.

intercommunal peace. This understanding, in turn, enables analysts to better determine intelligence requirements that will help them predict the incidence of communitarian strife.

Overall, the case studies show that the model is a good guide, but for it to be an effective tool, the analyst needs to put aside her own analytical preferences and accept the methodology. It takes discipline and rigor to do so. Area experts who used the methodology to produce the case studies displayed different degrees of acceptance. Some felt uncomfortable with the "straitjacket" the model imposed upon them or focused on the shortcomings of the model (described later in this chapter). Others came up with results that went against their intuitive assessments and wanted to "adjust" the model to "make it work." But the lesson of the experience to analysts who might use the model is this:

---

*Read the theoretical section carefully and address the questions (in the Appendix) faithfully. If the results differ from the expected findings, ask questions about what inputs might have led to the results that were expected. If the results are in rough agreement with the expected findings, use the model and its matrices to see what changes on the group or state side may lead to different results. The questioning process by itself should improve the analytical work.*

---

The model's data requirements produced different reactions among the various case study authors. Faced with an absence of data, some generalized from their knowledge of the country in question, others brought in data that was not directly pertinent, and still others acknowledged the lack of data and moved on after making certain assumptions for their estimates clear. Again, it took the effort of the editor to limit the inputs to only the data that were relevant as well as to make it clear when the case study author was making an estimate and when the data were available. The lesson of the experience to analysts using the model is this:

---

*The model generates substantial data requirements and forces a confrontation with data needs; it highlights not only what the requirements are but also their importance. Rather than becom-*

*ing bogged down in the data requirements, the analyst needs to keep in mind the goal, namely, collecting the information essential to make a determination of group and state capacities. The final analytical needs drive and prioritize the data collection; in itself, this should be useful for an analyst.*

The need to make assessments produced probably the most difficulty for the case study authors. Faced with a lack of a clear-cut answer in some areas, some case study authors initially refused to make an either/or choice. But this stymies the process and cannot be a solution. In cases of ambiguity, the case study authors were required to outline their reasons pro and con for each assessment; where it was possible, some of the case studies explored alternative outcomes to which the different assessments might have led. That is the intended purpose of the model: to force a closer analysis of what variables might cause potentially different results. The lesson of the experience to analysts using the model is this:

*When the choices are neither white nor black but gray, reexamine what might make them a lighter or a darker gray, and if the results are different (on the matrices), pay attention to the factors involved as potentially crucial variables.*

## BENEFITS OF USING THE MODEL

At a more specific level, the case studies illustrated that working with the model offers to analysts a number of specific benefits. The model can help area specialists structure a research design. It also helps introduce area experts to factors that are universally important in ethnic conflict. In other words, the model can help analysts hear through the noise that is part and parcel of intelligence analysis. Intelligence analysts are often presented with a bewildering array of data, some of which is incomplete and much of which is useless. The model can help them focus on what is important with regard to the potential incidence of ethnic conflict. For analysts who are accustomed to focusing on a particular country that has not suffered from communitarian strife in the past, the model offers guidance on which questions are important to ask. Saudi Arabia, for example, has not suffered from ethnic conflict, but the model suggests that if mobilization of the Shi'a takes place, violence is the likely outcome. Thus, the model increases the analyst's awareness of future events that merit close attention.

The model's tools for assessing closure offer a useful way to compare groups within a country in order to focus intelligence requirements on the most likely trouble spots.  Working with the model, analysts can be assured that they are integrating the tools and knowledge of years of comparative work into their assessments.  Of course, the model is no substitute for a comprehensive knowledge of the country in question, but it does facilitate attempts to structure thoughts and helps analysts avoid repeating mistakes made by students of other cases of ethnic conflict.  In countries that are not newcomers to communitarian strife, the framework also helps to explain the differences in how particular groups engage in political competition and how a regime will react to each of them.  For example, in Ethiopia, the Tigray-dominated regime is likely to deal in very different ways with challenges by the Amhara and the Oromo.  The model explains why the reaction is likely to be different.

In addition, the model explains the role of identity entrepreneurs and other key individuals, putting their efforts in a larger context. Too often, the role of leadership is ignored and conflict is explained entirely as a result of impersonal factors such as communal hatred, economic deprivation, or political disenfranchisement. Yet such factors, all too common throughout the world, do not always generate conflict. Violence requires individuals who are willing to take the risk and organize and lead groups. *Ethnic attachments or fears do not start violence, people do.*  The breakup of Yugoslavia, for example, cannot be understood without appreciating the role of Slobodan Milosevic in harnessing and deepening Serbian ethnic nationalism. Similarly, peace demands that individuals risk unpopularity and make concessions. The presence of such leaders, or their absence, explains a large part of the puzzle of why certain conflicts occur while others do not.  Nelson Mandela's strong leadership and his standing in South Africa was an important reason behind the largely peaceful transfer of power to majority rule.

Working with the model allows the integration of events that are normally viewed as distinct, such as the relationship between group grievances and communal mobilization. Politics in general is an integrated whole. To understand the impact of any particular event, it must be seen in the broad context of a country's political space. The change of government in Iran, for example, may decrease Saudi Shi'a resources, thus making Shi'a mobilization and potential conflict less likely. Similarly, the rebuilding of the Somali state may accentuate the Somali challenge to the regime within Ethiopia. The Ethiopian-Eritrean war has dampened—temporarily—the level of interethnic competition in Ethiopia. The model helps analysts bridge the gap between country-specific and regionwide events. It helps analysts identify important factors that might cause ethnic conflict and integrate them into one assessment.

Equally important, the model helps anticipate and explain peaceful as well as conflictual outcomes. Thus, analysts can anticipate smooth communal relations as well as the possibility of future conflict. This is far from a trivial point:

---

*Virtually every state in the world has some substate group differences; when compared to the universe of potential strife, the actual level of communitarian strife is extremely low. Peaceful relations between groups are the norm. Too many models of conflict overpredict the incidence of strife and underpredict the occurrence of cooperation.[2] Indeed, the greatest danger for analysts is probably the overprediction of communitarian violence.*

---

As the Saudi case shows, peace can occur when there are strong disincentives to group mobilization, such as when the regime in question is strong and leaves no doubt as to the price the mobilizing group will pay. Policymakers have a high level of interest in the political stability of some countries; some may find it reassuring that the Saudi Arabian monarchy is in no immediate danger from the Shi'a, even though the Shi'a population is disgruntled and faces a high degree of discrimination.

---

[2]James D. Fearon and David D. Laitin, "Explaining Interethnic Cooperation," *American Political Science Review*, Vol. 90, No. 4, December 1996, pp. 715–735.

Another strength of the model is that it allows analysts to vary conditions and data points and understand how these changes might affect the potential for ethnic conflict. As a result, unclear cases, or cases where events might go in different directions, can be better anticipated. Analysts often must make judgments on the basis of scarce data. The model highlights areas where data are scarce and illustrates how these gaps can greatly affect future predictions.

Analysts can use the model to create sets of future indicators for a conflict. Through the use of specific scenarios, even hypothetical situations (centering on different types of group mobilization) can be modeled. For example, in cases where ethnic groups are not mobilized, the analyst can examine the possibility of mobilization along ethnic lines, the potential pathways of mobilization, and their likely consequences. Such an exercise has a specific long-term warning use. Similarly, through the use of scenarios, the unfolding of group mobilization may be modeled even though not all of the elements of mobilization are yet in place. Such modeling can point to potential paths that, if identified early, may lead to warning and preparation. Thus, in the Ethiopia case an analyst can foresee one sort of future for the country if the regime-created parties prove genuine and gain popular support, namely that the potential for violence is less likely. But if the parties receive little support, a different future becomes more likely: conflict arising from ethnic mobilization by one of the major non-Tigrayan groups. Analysts can construct an array of indicators that will help them better determine when a country is moving toward conflict.

## CAUTIONARY NOTES ON USING THE MODEL

Because of its simplicity and emphasis on parsimony, the model has some limitations. The model shows indirectly the limits of the relative deprivation theory. The degree of closure is important in discerning the intensity of grievances, but it is also clear that the most deprived groups are not necessarily the ones that will mount a challenge to the regime. The Serbs in Yugoslavia did not face great discrimination. Their grievances stemmed primarily from their dominant myths of being a "martyr nation" and having a natural "right to rule" the country. The Ethiopian case showed that the Amhara believe that political power is theirs by right and are reluctant to share it with other groups.  In short, already powerful groups often seek a greater degree of closure and are willing to engage in violence to bring it about.  Dominant groups are particularly likely to engage in violence when their dominance is threatened.  The structure of closure is important in determining social stratification in a country, but the imagined grievances that stem from nationalist myths and self-perceptions and are exploitable by talented identity entrepreneurs can be just as powerful a catalyst for mobilization as real material grievances.  The important issue is the leadership's skillful mobilization and portrayal of grievances.  Thus, attention to the views and myths of dominant groups seems especially warranted. Analysts need a nuanced view of the structure of closure and should keep in mind that the most disadvantaged are not necessarily the most dangerous or the most likely to rebel.

The model's simple strategy choices contained in the matrices also need to be more refined.  For example, clever regimes often use several tactics simultaneously:  Carrots and sticks are not alternative strategies, where one is tried when the other fails. Rather, the two are used together as part of a greater whole. The "exploit" choice tries to capture this strategy, and it spans the range of "tough negotiating" to "implicit and/or selective repressive measures."  But a better understanding of conflict potential requires a more nuanced view of the range of choices in strategies of both the regime and the group.

Another aspect that the model may need to stress more is the role of uncertainty and fear in fomenting conflict.  The disintegration of Yugoslavia, for example, occurred in part due to the uncertainty that resulted from the power vacuum after Tito's death.  All of the Yugoslav groups became more concerned about their future security. Uncertainty can arise from the end of a particular regime type, a new social group arising, or other alterations that threaten to leave one or more groups subject to a greater degree of closure.  Under certain circumstances, all groups can fear an increase in closure even though this, of course, is impossible in reality.  When formulating intelligence requirements, analysts need to focus on the beliefs of groups and how secure they are in their social, economic, and political position.  The uncertainty and fear between groups are taken into account indirectly in the model (through the focus on drivers of mobilization), but they may need more explicit treatment.

Related to this problem is the nebulous but important question of the strength of ethnic identity. That aspect is captured partially in the potential of the group to mobilize. However, a more explicit focus on the intensity and strength of ethnic identity, as well as the presence of cross-cutting ties between communal groups (e.g., membership in the same labor union or allegiance to a shared political principle) as a way of offsetting communal differences and reducing perceptions of closure, may be warranted. For example, the relatively weak sense of Oromo identity in Ethiopia (at least in contrast to the Amhara) limits their potential for mobilization. Conversely, an active state policy of reducing differences between groups may turn out to be crucial. Thus, despite the extensive differences and rivalry between black organizations or groups in South Africa, a strong supragroup feeling of common ties among the black population of South Africa emerged, no doubt due to the specific state policy of treating all blacks as belonging to the same social class (thus decreasing communal differences).

This touches on the larger point that the role of communal groups as political actors is often more potential than real. Some "groups" may have little corporate existence until an "identity entrepreneur" mobilizes them ("conscientizes" them) or another event occurs that leads to greater group awareness. In the first stage of the model, the analyst should take an imaginative look at the identification of closure, in that she should think about the potential emergence of group identities based on patterns of, sometimes implicit, discrimination or privilege. The tendency is to focus on already existing groups. But the analyst should keep in mind that identity is fluid and apt to change, and that new "groups" can arise as a result of skilled mobilizational appeals. The rise of "Yugoslav" identity, which the consociational governing structure in Yugoslavia seemingly and paradoxically found so threatening, is a case in point. And regional differentiation can bring together rapidly into one "group" what may be a collection of diverse peoples.

Many, if not most, regimes pursue policies of "nation-building," usually understood as an attempt to homogenize the population of the state. There is nothing new about this policy, and contemporary "homogeneous" states like France (which pioneered this approach), Germany, or Italy became homogeneous only as a result of a century or more of often brutal centralization and state-sponsored "nation-building." The policy has been imitated in many developing countries after decolonization. Nation-building on one hand is inclusive, as it offers individuals access to power, status, and wealth. On the other hand, nation-building is exclusive, as it denies the value of cultural diversity and deprives traditional elites of their leading positions in society. Nation-building is a strategy that can change the very dynamic of the model by affecting group identification itself. Attention to its processes, its rate of success or failure and its consequences, should be a component of any analysis of potential for communitarian strife. The model takes this into account, but the point needs special attention in long-range forecasts.

The role of other states also may need to be brought into the model in a more direct fashion. The model explores how a foreign government can boost a mobilizing group's resource and organization base, and the case studies suggest that events abroad can have an even broader impact on mobilization through their example. For example, Saudi Shi'a ties to foreign groups have in the past inspired them to resist their government even when foreign governments have not provided any concrete aid. The success of violence in Lebanon, Iran, and elsewhere became a model for Shi'a throughout the world to emulate. Foreign aid can strengthen a state's fiscal and economic position, but it can also have a variety of effects on group mobilization. A group may fear being overwhelmed by a strengthened government, or it may fear a change in its social ranking. Conversely, foreign aid to the group may only strengthen the dominant group's perception of the challenging group as a "fifth column" and "traitors." Also, events across a border can affect a group's perception of closure and its propensity to mobilization (as a tipping event), particularly when it involves co-ethnics abroad. The model allows for the integration into the analysis of nonmaterial influences and assistance from abroad, but analysts need to make certain that such factors are considered adequately.

## ADDITIONAL USES OF THE MODEL

Besides guiding intelligence analysis, the model has uses as a tool for policy guidance. For example, it can be used to explore the effects of external armed intervention on intergroup relations. Support, even implicit, for some groups or factions by the intervening forces may cause a power shift among all the groups, causing some to be hostile and others to be supportive. The model provides a tool to assess the consequences of intervention.

This point touches on the usefulness of the model in directing policy. By using the model to test future outcomes, it can highlight the advantages and disadvantages of various policy alternatives. For example, if the model suggested that the creation of a more pluralistic system would improve a regime's accommodative capacity and lower the potential for violence in the face of incipient group mobilization, this could become an objective of U.S. foreign policy. Similarly, the United States might be able to reduce foreign meddling or otherwise prevent the mobilization of an ethnic group.

Particularly useful for policymakers would be the use of the model to determine which conflicts, if any, are especially difficult to resolve and when an external diplomatic intervention may be in order to stabilize the situation. When regimes are weak and exclusive, when closure is high, and when mobilizing events are common, neither the regime nor the communal group is likely to favor a peaceful solution. Indeed, both sides may consider any peace as a breathing space to rearm and mobilize in preparation for the next round of conflict. Knowledge of such a situation, and the ability to show in a causal fashion the reasons for coming to such a conclusion, would inform policy choices.

## FUTURE RESEARCH AND FURTHER DEVELOPMENT OF THE MODEL

As should be abundantly clear by now, this effort represents a first cut at the development of a model that pursues a different path from vulnerability analyses and which can be used by intelligence analysts for intensive assessments of a specific country's likelihood for communitarian strife. This effort is not the final word, and the authors recognize substantial areas for further development and refinement of the model.

One, the model focused on understanding one particular kind of ethnic action: the rise of an ethnic group challenging the state. It did not examine directly other kinds of ethnic competition, such as those that ensue when several ethnic groups compete with one another (with the state acting either as umpire, abettor, or participant). The model took such competition into account to a limited extent, but its focus is on the group-versus-state conflict. Dealing with pathways to conflict in multigroup situations may require other models based, to some degree or another, on the one offered here. Alternatively, it may be possible to "model" these alternative paths to strife as variant cases of the base model offered in this book. Indeed, as the case studies showed, it is possible to apply the model to situations that are more complex than the simple dyadic interaction. One way of proceeding is to group the potential challengers to the state into one "supergroup." If tensions among groups making up such a supergroup are likely, then the coding of leadership and popular support aspects of the group will be affected. For heuristic purposes, the use of the model in such a manner may suffice, but it is not a perfect tool by any means to deal with the more complex situations. In any event, modeling multigroup situations requires additional research.

Two, perhaps most important, the model can be developed beyond the simple strategy choices in the matrices. Just as the strategy choices stem from the assessment of the resources of the group and state against each other, it is feasible to develop more detailed pathways of action within these choices. For example, the "intimidate" strategy could be developed to span the range of the intensity and type of actions that a group could undertake. Based on a more detailed knowledge of the resource and support base of the group, it is theoretically possible to anticipate the type of violent action the group would be likely to embrace. These choices could number up to a dozen different categories and range from an armed uprising and a guerrilla war in the countryside to a select terrorist-style bombing campaign.[3] The same expansion of choices could be undertaken with the other strategy choices. This path of additional research also would throw light on the probable aims of the mobilized group

---

[3]At this stage, the model, or at least its matrices, would need to borrow from game theory and perhaps veer off into formal modeling. In terms of substantiation of the link between the choice of type of violence by groups as well as the state, depending on the resource structure and the type of opponent they are facing, there is a substantial literature on the topic. For some examples, see Stathis N. Kalyvas, "Wanton and Senseless? The Logic of Massacres in Algeria," *Rationality and Society*, Vol. 11, No. 3, pp. 243–285; and Leonard Wantchekon and Andrew Healy, "The 'Game' of Torture," *Journal of Conflict Resolution*, Vol. 43, No. 5, October 1999, pp. 596–609.

across the range of strategy choices. Because the set of potential choices would increase exponentially, the additional matrices would be complex and mathematically unwieldy and thus well-suited to development as a software program.

Finally, this research effort did not focus on disproving the theoretical claims offered here in the context of other alternatives that may exist in the literature. Such work too requires further consideration and effort.

# QUESTIONS AND GUIDELINES FOR THE ANALYST

This appendix derives the practical implications of the model for the analyst through the use of questions and guidelines to address aspects of each of the model's three stages. For the sake of clarity and in order to come as close as possible to specific indicators and essential elements of information, each set of questions is grouped and coded (up to four characters) with a series of letters and numbers. The coding sequence (explained below) aims to provide a step-by-step guide for an analyst in organizing the analysis. The guidelines are presented so as to guide the analyst in using the tables and matrices.

The first character in the coding sequence of all questions is a letter (capitalized, ranging from A to D) that pertains to the overall position of the question within the framework of the model. Thus, all questions pertaining to the first stage of the model, or the potential for strife, have codes starting with the letter "A"; all questions pertaining to the second stage of the model, or the mobilization of the group for political action (thus transforming the potential into likely strife), have codes starting with the letter "B"; all questions that aim to assess the state's capabilities, or the half-step of the third stage of the model that aims to establish the state's strengths and weaknesses in key areas, have codes starting with the letter "C"; all questions (really guidelines at this stage) to assess the strategic bargaining process have codes starting with the letter "D."

The second character in the coding sequence is a number that pertains to one of the subcategories in each of the three stages of the model. For all questions on the potential for strife (i.e., questions with a code starting with the letter A), the second character of the code is a number ranging from 1 to 4 that refers either to the political factors (A1), the economic factors (A2), the social factors (A3), or the overall assessment of potential for strife (A4). For all questions on the mobilization of the group (i.e., questions with a code starting with the letter B), the second character of the code is a number ranging from 1 to 6 that refers either to incipient changes (B1), galvanizing ("tipping") events (B2), leadership (B3), resources and organization (B4), the foreign element (B5), or the overall assessment of mobilization (B6). For all questions on the assessment of the state's strengths and weaknesses in key areas (i.e., questions with a code starting with the letter C), the second character of the code is a number ranging from 1 to 3 that refers either to the accommodative capability (C1), the fiscal and economic capability (C2), or the coercive capability (C3). For all guidelines to assess the strategic bargaining process (i.e., questions/guidelines with a code starting with the letter D), the second character is a number (either 1 or 2) that is related either to the measurement of the capacities of the group and the state (D1) or the comparison of the preferences within the bargaining process (D2).

In most of the categories, the breakdown into more specific questions or guidelines continues, and the codes reflect the further subcategorization. The third character is a letter (not capitalized, ranging from a to e) that further distinguishes the questions and their relationship to the larger categories. For example, within the overall category of assessment of the potential for strife ("A") and its subcategory of assessing closure in the political realm ("A1"), questions are further subdivided to address the distribution of power among the top political executive (A1a) and security (A1b) authorities.

The coding sequence also extends in some cases to a fourth character that further differentiates some questions or guidelines within the model framework. The fourth character is a number, ranging from 1 to 4. For example, the subcategory explained just above (A1a) is further divided into two groups, with questions coded A1a1 referring to indicators that focus on the highest-ranking individuals, and questions coded A1a2 referring to the indicators of the lower-ranking bureaucrats.

## A.  THE POTENTIAL FOR STRIFE

Assessing the potential for strife is especially applicable to countries that lack any visible and organized ethnic movements.  But the assessment is also useful for countries where such movements exist, because it allows the analyst to ascertain the depth of the ties that bind the existing groups as well as point to the potential for new groups to arise.  Both of the latter are especially useful for long-term assessment of the potential for strife.

As outlined in the model, the analysis must examine all three arenas of closure:  political, economic, and social.  The static "snapshot" of the current situation must be supplemented with a dynamic assess-ment of the ease of changing current stratification patterns.  But first the analyst needs to assemble data on the substate characteristics of the given population before proceeding to the specific questions.  The task involves compiling data on racial, religious, linguistic (including significant dialects), regional, or other significant cultural differences.  For populations of small countries, where kinship ties may remain important, extended clan groups may also need to be examined.  The purpose of the compilation is to establish the "terms of reference" for examining closure boundaries.  At this point, the task is simply to note the numerical data and distribution patterns of the given subpopulation.  The analyst should note discrepancies between figures from different sources (for example, for the popula-tion size of one linguistic group, members of that group and the gov-ernment may provide quite different numbers).

The following outline presents the topics of the questions that an analyst needs to consider in assessing the potential for strife.  The outline is presented to make clear the organization of the section and the logical connections between the questions.  The actual questions follow the outline.

## Outline of Questions for Assessing:
## A. The Potential for Strife

### A1. ASSESSING CLOSURE IN THE POLITICAL REALM

**A1a.** Top political authorities

    *A1a1. Highest-ranking individuals*

    *A1a2. Director-level managers*

**A1b.** Top security authorities

    *A1b1. Highest-ranking individuals*

    *A1b2. Director-level managers*

**A1c.** Assessment of static factors at the political level

    *A1c1. Comparing patterns at the political executive level*

    *A1c2. Comparing patterns at the security level*

**A1d.** Assessment of dynamic factors at the political level

    *A1d1. General framework for change*

    *A1d2. Framework for change by specific groups*

### A2. ASSESSING CLOSURE IN THE ECONOMIC REALM

**A2a.** Distribution of wealth

    *A2a1. The elite level*

    *A2a2. The general population level*

**A2b.** Assessment of static factors at the economic level

**A2c.** Assessment of dynamic factors at the economic level

    *A2c1. General framework for change*

    *A2c2. Framework for change by specific groups*

### A3. ASSESSING CLOSURE IN THE SOCIAL REALM

**A3a.** Status distribution: static aspects

    *A3a1. Determinants of status distribution*

    *A3a2. Benefits and restrictions based on status*

**A3b.** Status distribution: dynamic aspects

    *A3b1. General framework for change*

    *A3b2. Framework for change by specific groups*

## A4. OVERALL ASSESSMENT OF CLOSURE

**A4a.** Framework for closure

**A4a1.** *Patterns of privilege and deprivation*

**A4a2.** *Degree of relative deprivation*

**A4a3.** *Potential for change*

**A4b.** Ranking the groups

# A1. Assessing Closure in the Political Realm

An examination of the political realm consists of looking at the static and dynamic aspects of the distribution of power among the political executive and security authorities. Using the terms of reference already compiled (data on population characteristics), the analyst should examine the distribution of power among the top political executive (questions coded A1a) and security (questions coded A1b) authorities, with both the uppermost and more broadly understood upper-level officials scrutinized. The actual characteristics exhibited should then be compared with the characteristics of the population (guidelines for comparison are coded A1c). The questions and guidelines coded A1a–A1c consist of the static factors. A final element follows, the evaluation of dynamic factors in assessing closure in the political realm (questions coded A1d). The following questions should guide the examination.

**A1a. Top political authorities.** The top political authorities need to be examined at two levels: the highest-ranking individuals (questions coded A1a1) and the director-level managers (questions coded A1a2) who make the everyday decisions and implement government policies. An examination of both levels is necessary so as to avoid skewed results in either area.

*A1a1. Highest-ranking individuals.* Which substate characteristics[1] do the top five to ten individuals in positions of executive political authority exhibit? The top executive political positions are those where the final responsibility for the implementation of gov-

---

[1] The "substate characteristics" are defined in the paragraph under the previous heading "A. The Potential for Strife."

ernment policy rests.  Generally, these positions include (though may not be limited to):  the president, the prime minister, and the ministers concerned with the economy (treasury, finance, industry, energy, etc.) and internal administration (interior, justice).[2]  Top executive personnel in the security realm are the subject of another question (A1b1).  In cases of divided government (real constraints on executive power), the legislative and/or judicial branches also may need to be examined in the manner presented above.  In states where provincial administration has substantial political power (as in federal states), the key figures for the province also need to be assessed.

*A1a2. Director-level managers.*  Which substate characteristics do the individuals in charge of everyday implementation of governmental policy exhibit?  The positions in question here are sectional chiefs of the executive bureaucracy (one or two grades down from the ministerial level).  The data pertain to the same ministries in the economy and security realms identified in the preceding question (A1a1).  The same qualification as in A1a1 applies to states with divided governments and to states where the provinces have substantial room for independent action.

**A1b.  Top security authorities.**  The assessment of top security authorities proceeds in a fashion similar to the examination of the top political authorities, with one assessment at the level of the highest-ranking individuals (questions coded A1b1) and a complementary assessment at the director level (questions coded A1b2).  The rationale for the two levels is the same as in the political level assessment: to avoid skewed results in either area.

*A1b1. Highest-ranking individuals.*  Which substate characteristics do the top five to ten individuals in positions of authority in the security sphere exhibit?  The top positions in the security apparatus are those where the final responsibility for the implementation of governmental policy rests.  The security apparatus consists of (but is

---

[2]In some cases, the most important executive authority may not have any formal state title but instead go by "Great Leader" or "Supreme Guardian."  Since the aim of the exercise is to focus on the top leaders, the analyst should go by the function performed rather than by formal title.  In most cases, formal titles are the best tip-off for the function performed.

not necessarily limited to) the following:  military, police, intelligence, foreign relations.  The positions in question are the ministers and chiefs of staff for the various apparati.  In states where the provincial authorities have substantial room for independent action (and have their own security structure independent of direct central control), the examination also needs to be conducted at the provincial level.

*A1b2.  Director-level managers.*  Which substate characteristics do the individuals who form the upper levels of the security apparati exhibit?  The positions in question here are the upper-level ranks of the security bureaucracies identified in the preceding question (A1b1).  For example, concerning the military, the ranks of the officer corps at the level of colonel and above, form the group for examination.  The same qualification as in A1b1 applies to states with shared central-provincial authority.

**A1c.  Assessment of static factors at the political level.**  The data assembled in A1a and A1b present a picture of the top authorities, but they mean little without a comparison to the situation existing in the society as a whole.  Any patterns that may point to closure emerge only in such a comparison.  The questions below are meant to guide the comparison, with one set of questions (coded A1c1) focusing on the political executive authorities and another (coded A1c2) looking at the security authorities.

*A1c1.  Comparing patterns at the political executive level.*  What is the discrepancy at the political (executive) authority level (A1a), both in terms of the elite (A1a1) and upper-level administrators (A1a2), between the actual characteristics exhibited and the characteristics of the population as a whole?  Discrepancies between the actual and proportional characteristics may indicate limitations on access to power (closure).  Rather than focusing on small differences, the analyst should note major differences in the two patterns (in over- and underrepresentation).  For example, compare the population of country X and its political (executive) elite and top administrators in terms of linguistic preferences.  Assuming that 35 percent of the country's population are native speakers of language z, the fact that 5 percent of the elite and 10 percent of the top administrators are native speakers of language z may indicate closure in the political realm (with native speakers of language z in an inferior position).  In case of a comparison of the actual and proportional figures at the provincial level, the analyst should assemble data and compare both characteristics at the provincial (rather than state) level.

*A1c2. Comparing patterns at the security level.*  What is the discrepancy in the security sphere (A1b), both in terms of the elite (A1b1) and upper-level ranks of the apparati (A1b2), between the actual characteristics exhibited and the characteristics of the population as a whole?  The same guidelines outlined in A1c1 apply, in that large discrepancies (either over- or underrepresentation) between the actual and proportional characteristics may indicate limitations on access to power (closure).  When provincial-level characteristics are also to be examined, the same guidelines as in A1c1 apply.

**A1d. Assessment of dynamic factors at the political level.**  The static snapshot of possible closure that is determined in A1a–A1c needs to be supplemented by an examination of the restrictions on access to positions of power in the political and security spheres.  The dynamic component is indispensable to a long-range assessment, for it provides grounds for establishing the possibility for changing the situation identified in A1a–A1c.  Formal restrictions should be the primary focus (because of the ease in identifying them), but any informal restrictions also should be noted.  Both the ease of change at all (questions coded A1d1) and ease of change on the part of some substate groups (questions coded A1d2) need to be examined.

*A1d1. General framework for change.*  Is suppression of overt dissent a part of the governing process?  Is there a codified process for the legitimization and/or replacement of political elites?  These questions assess the overall state of development of the political institutions and establish whether change of access to the political realm by peaceful means can be contemplated realistically.

*A1d2. Framework for change by specific groups.*  Are there any restrictions on access to the political realm (voting rights, ability to hold office) along the lines of characteristics of the population at the substate level?  Are there any restrictions at the level of substate characteristics on redress of grievances in the judicial system?  Are there any restrictions on assembly along the lines of substate characteristics? Are there any restrictions on, or are any groups singled out for, service in the security apparati? Formal rules curtailing access are easily identifiable and may be found in constitutions and specific laws. Informal rules curtailing access are not as easy to identify, but in conjunction with the answers to economic (A2) and social (A3) closure indicators, they may become clear. For example, formal restrictions on access to politics may be written so as to target specific economically disadvantaged substate groups without naming those groups explicitly.

## A2. Assessing Closure in the Economic Realm

Using the same terms of reference as in A1 (data on population characteristics), the analyst should examine the distribution of wealth in the society (questions coded A2a) and then compare the actual characteristics with the characteristics of the population (questions coded A2b). The resulting assessment comprises the static snapshot of the situation. Another set of questions (coded A2c) evaluates the dynamic factors in assessing closure in the economic realm. The following questions should guide the examination.

**A2a. Distribution of wealth.** An analysis of the distribution of wealth in a given society needs to proceed at two levels: the patterns at the elite level (questions coded A2a1) and the patterns at the level of the general population (questions coded A2a2). An examination of both levels is necessary so as to avoid skewed results in either area.

*A2a1. The elite level.* Which substate characteristics does the wealthiest 1 percent of the population (identified in terms of assets) exhibit? The 1 percent figure is adopted for convenience only, as a somewhat arbitrary way to identify the wealthiest; alternatively, figures anywhere from 2 to 5 percent also may provide the information. The central aspect of this analysis is to identify the richest stratum of the population. Data may be difficult to obtain on this point (because of holdings in other countries), and there is an acute need for information here beyond what may be officially supplied.

*A2a2. The general population level.* What are the patterns of wealth (in terms of substate characteristics) in the population as a whole? A standard measure, such as per-capita income distribution, needs to be applied. The central aspect of this analysis is to identify the prevailing patterns of wealth in the population as a whole. Accurate data for this analysis may be difficult to obtain.

**A2b. Assessment of static factors at the economic level.** By itself, the data assembled in A2a provide a picture of the patterns of distribution of wealth, but this means little without a comparison to the breakdown of the society into substate groups according to the terms of reference gathered initially. The specific questions for the analyst are: What is the discrepancy in the pattern of wealth distribution, both in terms of the elite (A2a1) and the general population (A2a2), between the actual characteristics exhibited and the characteristics of the population as a whole? Discrepancies between the actual and proportional characteristics may indicate limitations on access to wealth (closure). The analyst should note major differences in the patterns observed (in over- and underrepresentation). For example, compare the population of country Y and its wealthiest 1 percent and the general pattern of wealth distribution among the population in terms of religious preferences. Assume that 60 percent of the population of the country are adherents of religion x. The fact that 95 percent of the wealthiest 1 percent are adherents of religion x and that 80 percent of the wealth is concentrated in the hands of adherents of religion x may indicate closure in the economic realm.

**A2c. Assessment of dynamic factors at the economic level.** The static snapshot of possible closure that is determined in A2a–A2b needs to be supplemented by an examination of the restrictions on access to wealth. The dynamic component is indispensable to a long-range assessment, for it provides grounds for establishing the possibility for changing the situation identified above. Formal restrictions should be the primary focus (because of the ease in identifying them), but any informal restrictions also should be noted. Both the ease of change at all (A2c1) and ease of change on the part of some substate groups (A2c2) need to be examined.

*A2c1. General framework for change.* Is suppression of independent economic activity a part of the governing process? Are there any fundamental noneconomic barriers to activity leading to capital accumulation? These questions assess the overall state of the economic institutions and establish whether change of access to wealth through economic means can be contemplated realistically.

*A2c2. Framework for change by specific groups.* Are there any restrictions on access to wealth (property laws, tax laws, limits on establishing firms) along the lines of characteristics of the population at the substate level? Are there any restrictions on consumption that may disadvantage individuals sharing certain substate characteristics? Formal rules curtailing access are easily identifiable and may be found in specific laws. Informal rules curtailing access are not as easy to identify, but in conjunction with the results of answers to political (A1) and social (A3) closure indicators, they may become clear. For example, formal restrictions on access to wealth may be written so as to target specific economically disadvantaged substate groups without naming those groups explicitly. Examples of such practices may be discerned through attention to the tax structure, subsidies, trading patterns favored, and land transfer practices. For example, tax preferences and/or subsidies may exist for certain industries (and these industries may be identified with specific substate groups). In addition, certain economic activities may be defined as criminal and prosecuted. The analyst should note the informal restrictions that amount to limits on capital accumulation for some and the retention of wealth for those who are already wealthy.

## A3. Assessing Closure in the Social Realm

Using the same terms of reference (data on population characteristics), the analyst should examine the distribution of status in the society (questions coded A3a) and the possibility for changing the pattern of status distribution (questions coded A3b). The former provides a static snapshot of the possible closure in the social realm, while the latter provides the dynamic element. Differentiation of substate groups according to status may help pinpoint some informal closure rules in the political and economic realms. The following questions should guide the examination.

**A3a. Status distribution: static aspects.** Because status according to substate characteristics is, by definition, accorded to the entire group, this set of questions lacks the division into two levels of analysis (elite and mass) that was necessary in the assessments at the political and economic levels. Ascertaining status distribution should be carried out by first looking at the existing pattern of distribution (questions coded A3a1) and the specific benefits conferred or restrictions imposed on individuals belonging to a higher or a lower status group, respectively (questions coded A3a2).

*A3a1. Determinants of status distribution.* What is the basis for the pattern of status hierarchy and stratification in the state (vocational, hereditary, or hierocratic)? If vocational, are there any vocations identified with specific substate groups? If hierocratic, are there any groups identified specifically with the state (is the state defined in national/group terms or not)? For example, does the constitution speak of the state as an expression of a specific national group, or does it speak of the citizens of the state (not defined in national terms)? Are there any groups seen as foreign or unassimilated? Based on status differences, the analyst should put together a status stratification map of the society, ranging from the privileged to the pariah groups.

*A3a2. Benefits and restrictions based on status.* What kind of benefits accrue to members of the higher status groups simply on the basis of belonging to the group? Conversely, what kind of restraints affect members of the lower status groups simply on the basis of belonging to the group? For example, does belonging to a higher status group carry exclusionary rights to participate in the political realm, or is it simply a symbolic reference that no longer carries any tangible benefits? Similarly, does belonging to a lower status group carry any identifiable constraints on the range of economic activity in which an individual can take part? Are certain linguistic groups disadvantaged in state structures (for example, in the judiciary system)?

**A3b. Status distribution: dynamic aspects.** The static snapshot of possible closure that is determined in A3a needs to be supplemented by an assessment of the rigidity of the status stratification and the extent of mobility among the status groups. The dynamic component is indispensable to a long-range assessment, for it provides grounds for establishing the possibility for changing the situation identified in A3a. Formal restrictions on membership in status groups should be the primary focus (because of the ease in identifying them), but any informal restrictions should also be noted. Both the ease of change at all (questions coded A3b1) and ease of change on the part of some substate groups (questions coded A3b2) need to be examined.

*A3b1. General framework for change.* How fluid is membership in a group, and is mobility between status groups possible (or are groups defined so as to preclude movement between them)? Does the state establish rules on status groups, and does it punish those who do not follow them? Is there a norm (perhaps upheld by the state) of social tolerance, or is there a norm of exclusion and separation? These questions assess the overall possibility of whether change toward belonging to a higher status group can be realistically contemplated.

*A3b2. Framework for change by specific groups.* If mobility is possible, how easily is it accomplished? Are members of some groups precluded from mobility? What is the length of time in which an individual can change status groups? Is it a matter of rapid accomplishment in a given vocation, or is it tied to intermarriage and generations of residence in the same locality? Are certain groups especially disadvantaged by the existing rules on mobility?

## A4. Overall Assessment of Closure

The examination of the specific realms of closure performed on the basis of the questions in A1–A3 has described the cleavages and potential areas of closure within the society being assessed. On the basis of the assembled information, the analyst needs to put together an assessment of the rankings of the various groups in the society. First, the analyst should establish the framework for the rankings (questions coded A4a) and then put together an explicit ranking system along the lines of discovered closure (guidelines coded A4b).

**A4a. Framework for closure.** The composite assessment of the framework for closure needs to make use of the information gathered so far in a manner that is most useful to the overall model. In this sense, the data gathered so far should be reinterpreted along the following lines. Using the data compiled on the characteristics of the society, which groups (if any) are privileged and which are not? (Questions coded A4a1 address this issue.) Of the groups that are not privileged, what is the degree of relative deprivation? (Questions coded A4a2 address this point.) How easy is it to change the existing inequalities and imbalances? (Questions coded A4a3 address this issue.)

*A4a1. Patterns of privilege and deprivation.* Do any groups show a consistent pattern of privilege or deprivation across the three realms of closure? The analyst should make the assessment on the basis of the existence of potential closure patterns established in A1c1–2, A2b, and A3a1, treating the political realm (A1c1–2) as most important. In cases where strong patterns of closure exist in all three realms, a given group may be said to be experiencing a high degree of closure (either in a position of domination or dominated). In cases where less clear-cut patterns of closure exist or if the patterns do not extend across all three realms, a given group may be said to be experiencing a moderate degree of closure. If no clear pattern emerges, then closure along that specific variable may not be a factor in the given polity.

*A4a2. Degree of relative deprivation.* Assuming an identified high or moderate degree of closure, what is the degree of deprivation that a given group may be suffering? The analyst should make the assessment on the basis of the extent of potential closure patterns established in A1c1–2, A2b, and A3a2. Taking the most privileged group as the point of reference, how much are other groups deprived relative to it? In cases where far-reaching differences exist in all three realms, a given group may be said to be experiencing a high level of deprivation as a result of its closure. In cases where the differences are not clear cut across the three realms, a given group may be said to be experiencing a moderate level of deprivation as a result of its closure. In cases where differences are small, then deprivation is low.

*A4a3. Potential for change.* Looking at the groups experiencing a high or moderate degree of closure, how amenable is the existing situation to change? The analyst should make the assessment on the basis of the dynamic factors identified in A1d, A2c, and A3b. In cases where the situation is not amenable to change at all or is amenable to change only in an extremely lengthy (generational) fashion, a given group may be said to be experiencing an extremely rigid form of closure (either in a position of domination or dominated). In cases where the situation is amenable to change in an individual's lifetime, a given group may be said to be experiencing a moderately rigid form of closure. In cases where the situation is amenable to rapid change (a manner of years), a given group may be said to be experiencing a fluid form of closure.

**A4b. Ranking the groups.** On the basis of the assessments in A4a, the analyst should place the various groups, ranging from those in a position of privilege to those in a position of lacking privilege, using the levels of deprivation and rigidity to differentiate the various groups. The ranking represents a final assessment of the potential for certain groups to be willing to take up arms in order to change their position in the society. Rather than a strictly hierarchical ranking, some groups are likely to be advantaged in some areas and facing deprivation in others. The analyst must take such nuances into account. The assessment represents the final results of the application of the first stage of the model to a given case.

## B. TRANSFORMING POTENTIAL INTO LIKELY STRIFE

As outlined in the model, the assessments derived in the preceding section (questions coded A) only point to groups that are aggrieved and might be willing to contemplate violence in order to change the existing patterns of power and domination. To assess the likelihood of violence, the aggrieved groups must be analyzed in terms of mobilization potential.

There are five catalysts of mobilization that may lead to effective group action, and they need to be examined in detail. The questions below address each of the factors: incipient changes (questions coded B1), galvanizing events (questions coded B2), leadership (questions coded B3), resources and organization (questions coded B4), and the foreign element (questions coded B5). In addition, a final set of guidelines (coded B6) assesses the overall mobilization potential.

Whereas the nature of closure made a systematic examination of its components fairly straightforward, an examination of the factors needed for mobilization lacks such a uniform structure for two reasons. First, the time frame is different; unlike an examination of the structure and the persistent features of the society, an examination of mobilization processes has to take into account that mobilization can be quite rapid and surprising, even to the participants. Second, the specific nature of how the five factors identified as necessary for mobilization will demonstrate themselves is exceedingly difficult to anticipate. However, thinking in terms of the five categories can shape the analyst's understanding of the mobilization and its strength, provide an early understanding that what is happening is indeed mobilization along ethnic or communitarian lines, and is crucial to understanding the third stage of the model—the bargaining process.

The following outline presents the topics of the questions that an analyst needs to consider in assessing the transformation of potential into likely strife. The outline is presented to make clear the organization of the section and the logical connections between the questions. The actual questions follow the outline.

### Outline of Questions for Assessing:
### B. Transforming Potential Strife into Likely Strife

**B1. INCIPIENT CHANGES**

> **B1a.** Long-term changes

> **B1b.** Sudden changes

**B2. GALVANIZING ("TIPPING") EVENTS**

**B3. LEADERSHIP**

**B4. RESOURCES AND ORGANIZATION**

> **B4a.** Resources

>> **B4a1.** *Overall availability*

>> **B4a2.** *Mechanisms for extraction of resources*

> **B4b.** Organization

> **B4c.** Overall assessment

**B5. FOREIGN ELEMENT**

> **B5a.** Co-ethnics abroad

> **B5b.** International disputes

**B6. OVERALL ASSESSMENT OF MOBILIZATION**

## B1. Incipient Changes

An incipient change in closure and power relations can stem either from long-term trends (questions coded B1a) or sudden events (questions coded B1b). Ongoing small changes in a society are difficult to discern precisely because they are constant and ever present. However, long-term incremental change amounts to a fundamental change, and at some point, through a specific event, the full magnitude of the long-term changes becomes evident in a manner that upsets the existing power balance. Sudden events, such as technological breakthroughs or mineral discoveries in a certain region, also threaten the existing power balance. It is possible to predict the eventual effect of long-term changes, but the changes themselves are difficult for an analyst to discern. The sudden changes are easily noted, but they are impossible to predict because they are event-specific and, by definition, unexpected. Whatever the cause, the crucial aspect here is the change in power relations and the upsetting of the status quo through a specific event.

**B1a. Long-term changes.** Using the terms of reference ascertained earlier, what are the demographic, economic, and environmental trends in a given society? At what point will these trends amount to pressures for systemic change? What venues are likely for the actualization of such pressures? For example, if two substate groups inhabit a given state and the smaller group (one that is in a state of relative deprivation vis-à-vis the larger group) has a natural growth rate twice that of the privileged group, at some point the deprived group will become numerically larger. At that point, demands for recognition of the group's majority status (and redressing the group's deprived status) through political measures might be forthcoming. When the time is close to such a demographic shift, all the actors will understand that an incipient change is coming. The main task for the analyst is to note the point at which long-term incremental changes amount to a basic change in power relations.

**B1b. Sudden changes.** Using the earlier defined terms of reference, which groups are likely to gain and/or to lose from unexpected major breakthroughs or discoveries? For example, if major gold or oil deposits are discovered in a province populated primarily by a group that is in a subordinate position in that country, the expectation of a sudden inflow of capital into the province may amount to an unexpected shift in power balance in the country as a whole. Unlike B1a, where the trends are possible to predict, the analyst should look at the issue of sudden changes from the perspective of monitoring ongoing events and ascertaining their impact on the larger power relations and closure in a given society. In other words, an analyst should consider specific events from the standpoint of ascertaining their potential for causing disruptions to the status quo and closure.

## B2.  Galvanizing ("Tipping") Events

A public event that galvanizes group sensibilities can range from the firing of an outspoken newspaper editor to a violent crackdown on the members of a given group. As such, the specific form that a galvanizing event might take is nearly impossible to anticipate. But through a series of questions the analyst might at least anticipate that certain events may have the potential to become galvanizing events. Keeping in mind that the central point of a galvanizing event is the symbolism behind the action, the following questions may help in narrowing the range of possibilities and should be useful in structuring the analysis. The questions focus on group anniversaries, places of special significance, prominent personalities, and group-specific cultural traits.

Are there any regularly scheduled marches or demonstrations (related, for example, to a date that holds special group-specific meaning) that may lead to a clash? Is there a specific place of symbolic significance for the group (for example, a temple or shrine) that may be subject to destruction? Is there a specific figure whose death or imprisonment will lead to his or her martyrdom? Are there any specific physical or cultural characteristics of the group (headdress, facial hair, language, certain food or drink) that may be subject to outlawing or persecution by the state? As noted above, the list is almost inexhaustible because it is event-specific, but the four distinctive categories given above provide some structure in identifying the range of choices.

## B3.  Leadership

Just as in the case of galvanizing events, the emergence and background of leaders is event-specific and idiosyncratic. However, similarly to galvanizing events, a skeleton analytical structure can be put together so as to identify likely leaders. One realm of difference is related to the type of social structures in the given society. Therefore, a basic question is whether traditional (patrimonial) or bureaucratic power structures are paramount in the given society. Individuals emerging from organizations that are respected in the given society will be more likely to command respect and loyalty. In societies where patrimonial structures exist, some individuals assume a position of certain stature on the basis of birth and family. In societies where bureaucratic structures exist, some individuals who rise to a high position within them command a certain degree of stature. However, to be effective and act as a motivating force for the group, a leader must have a certain degree of charisma and be willing to take some risk to his safety. Keeping in mind the natural division between traditional and modern power structures that result in different types of respect and stature depending on the society, the following questions may narrow the choices in identifying potential leaders.

Are there any activists who disregard personal safety and comfort to energize the group? Are there any individuals whose speeches have a markedly and noticeably strong effect upon the audience? Are there any unusually talented individuals from a subordinate group who have risen quickly through the ranks in the armed forces or civil service? Are there leadership vehicles in place (unions, clubs, societies, etc.) so that a leader may emerge?

## B4. Resources and Organization

Unlike leadership and galvanizing events, it is easier to make an assessment of resources (questions coded B4a) and level of organization (questions coded B4b) of a given group. An overall composite assessment should follow (guidelines coded B4c). Since the existence of traditional (patrimonial) or modern (bureaucratic) power structures in the society affects the existence of resources and organization, the answer to that question (from answers regarding leadership issues, or B3) should be kept in mind. The following questions should assist in the analysis of the level of resources.

**B4a. Resources.** Resources that theoretically might be made available in conditions of full mobilization should be the goal of the assessment (questions coded B4a1), modified by an evaluation of the mechanisms for achieving full mobilization (questions coded B4a2).

**B4a1. Overall availability.** What are the overall levels of wealth of the group? Are there any other groups that might sympathize enough with the mobilizing group so as to make available some of their resources? Are there any specific wealthy individuals (either from the group or from outside of it) who sympathize with the group and/or its leader?

**B4a2. Mechanisms for extraction of resources.** Since the higher the deprivation the more likely the willingness to contribute resources, how high is the level and rigidity of deprivation of the group (also see A2b–c)? How effective a mechanism can the group construct to ensure compliance with "taxes for the group?" Does it need to rely on force for compliance, or can it rely on voluntary contributions?

**B4b. Organization.** The initial question on organization is whether the leader brings in an already existing organization to the mobilization process, for example as a result of a patrimonial basis to leadership (B3). The analyst should then attempt an assessment of the potential for the formation of a functioning organization on the basis of background and experiences of the group and its leader.

Does the leadership have first-hand knowledge of and experience with complex hierarchical organizations? Are there any individuals closely associated with the leader who have such knowledge or experience? What are the existing organizational mechanisms within the group that might be used by the mobilizing leader?

**B4c. Overall assessment.** The analyst then needs to link the answers about resources and organization (B4a–b) to assess the level of organizational effectiveness that the mobilizing group can achieve. If the leadership is in a strong position to mobilize and organize the resources of the given group, then it can be said to have high organization potential. If the leadership does not appeal to the whole group and/or is unlikely to have the organizational skills needed to utilize the resources effectively, then it can be said to have moderate organizational potential. If the leadership is weak, has limited appeal to the group as a whole, and has few of the skills needed to use the resources, then it can be said to have low organizational potential.

## B5. Foreign Element

Foreign assistance is a means of boosting the mobilizing group's resource and organization base. The assistance can be distinguished along two lines: group affinity (questions coded B5a) and independent state interests (questions coded B5b).

**B5a. Co-ethnics abroad.** Group affinity relates to the possibility that the members of the given group inhabit areas of neighboring countries as well as form distinct diasporas as a result of immigration. The questions for the analyst focus on both types of groups. Do the state boundaries cut across a given group's homelands? Are there any identifiable irredentist organizations in the neighboring country? What kind of resources and skills can the ethnic diaspora add to the resource base of the mobilizing group? Are there identifiable mechanisms in the diaspora that can channel the resources and skills back to the group?

**B5b. International disputes.** Do any of the neighboring states have existing disputes with the given state? How strong are the disputes? Under what conditions might the neighboring state give support to the mobilizing group? What kind of support (in terms of resources and organization) can the neighboring state offer?

## B6. Overall Assessment of Mobilization

As noted above, the criteria for assessing the potential for mobilization have to remain vague because of its event-specific nature. However, the analyst should draw certain scenarios designed to elucidate the range of possibilities in resources and organization, based on (1) projection of an incipient change, (2) assumption of a galvanizing event, (3) potential emergence of three types of leader (strong, moderate, and weak), and (4) foreign support (if applicable). The drawing out of scenarios should make clear the possible consequences of mobilization. In some cases, even a strong leader might not be able to command significant resources. In other cases, even a weak leader might have substantial resources at his or her disposal. The crucial role of the leader is evident in this type of assessment of mobilization potential.

The assessment represents a way to portray the effect that an aggrieved and subordinate group might have through mobilization. Ideally, in conditions where several aggrieved groups have been identified in stage 1, the analyst should examine the mobilization potential of all of them, ranking them in terms of strength of mobilization. Just as in the final assessment in stage 1 of the model, the analyst should place the various aggrieved groups on a continuum, differentiating them by the level of mobilization potential. The differentiation represents a final assessment of the likelihood that certain groups might take up arms in order to redress their position in the society. The assessment also represents the final results of the application of stage 2 of the model to a given case.

## C. ASSESSING THE STATE

As outlined in the model, the mobilization of a group for political action does not necessarily lead to violence; the interaction between the state and the mobilized group may lead to any number of outcomes, ranging from peaceful reconciliation to an attempt at genocide. Anticipating what the potential outcome might be is the focus of the third stage of the model, which deals with the bargaining process.

The analysis first needs to establish the state's capabilities in three categories: accommodative (questions coded C1), fiscal and economic (questions coded C2), and coercive (questions coded C3). The accommodative capability is most important, since it pertains to the central issue of domination and monopolization of power. The results then need to be compared to the group's capabilities so as to arrive at a determination of the strategies the two actors will pursue. Establishing the state's capabilities is a step analogous to the analysis performed in stage 2 (questions coded B). The actual matching up of resources and elaboration of expected outcomes is a more mechanistic process, carried out with the aid of tables and matrices (guidelines coded D).

The following outline presents the topics of the questions that an analyst needs to consider in assessing the state. The outline is presented to make clear the organization of the section and the logical connections between the questions. The actual questions follow the outline.

## Outline of Questions for Assessing:
## C. The State

### C1. ACCOMMODATIVE CAPABILITY

**C1a.** Responsiveness of political structures

*C1a1.* Level of inclusiveness

*C1a2.* Potential for change in political structures

*C1a3.* Overall assessment of responsiveness

**C1b.** Prevailing norms of governance

**C1c.** Cohesion among ruling elites

**C1d.** Composite assessment of accommodative capability

### C2. FISCAL AND ECONOMIC CAPABILITY

**C2a.** Fiscal health

**C2b.** Resource extraction potential

**C2c.** Size and wealth of ruling elites

**C2d.** Composite assessment of fiscal and economic capability

### C3. COERCIVE CAPABILITY

**C3a.** Command and control over the apparati of violence

**C3b.** Composition of the apparati of violence

**C3c.** Norms toward domestic use of violence

**C3d.** Suitability of the force for domestic use

**C3e.** Composite assessment of coercive capability

## C1. Accommodative Capability

Three categories of questions address the state's accommodative capabilities. The first category aims to ascertain the level of responsiveness of the state's political structures to popular will (questions coded C1a). The second category deals with the prevailing norms of governance that the political structures embody (questions coded C1b). The third category addresses the extent of cohesion among the ruling political elites (questions coded C1c). An overall determination of the accommodative capability is the final step (guidelines coded C1d).

**C1a. Responsiveness of political structures.** Assessing the responsiveness of political structures to popular will can be answered through two sets of questions. The first set addresses the level of inclusiveness of the political structures (questions coded C1a1). The second set deals with the possibility and ease of change of the political structures (questions coded C1a2). The result is a composite assessment of responsiveness (guidelines coded C1a3).

*C1a1. Level of inclusiveness.* Are the political structures inclusive or exclusive? Are there mechanisms in place (such as elections) that regularly hold political authorities accountable to popular will? Are there differing levels of inclusion/exclusion from the political process according to some group criteria? Are there certain groups that are habitually excluded from the political process on whatever basis, formally or informally? Constitutions and other laws are the specific indicators the analyst should consult. But the effort should go beyond the formal to the informal rules governing participation; these can be gauged from actual participation rates. The analysis should focus on evidence of low participation by a group as a potential indicator of informal exclusionary rules (some of the data gathered by the analyst while assessing closure—questions coded A— may be useful also at this stage).

*C1a2. Potential for change in political structures.* What are the mechanisms of conflict resolution built into the political structure? Are there established channels for change in political institutions? Again, the indicators are formal rules and laws, but the actual evidence of whether these channels are effective needs to be assessed by whether changes in institutions actually occur over time. Lack of changes over a lengthy time may point to possible inflexibility of the political structures in practice even if formal rules make them appear flexible.

*C1a3. Overall assessment of responsiveness.* On the basis of C1a1 and C1a2, the analyst needs to make an assessment of the overall level and extent of responsiveness of the political institutions. The level of inclusiveness and the flexibility of the political structures deserve equal weight.

**C1b. Prevailing norms of governance.** Are there tolerant norms in place in the society, and are they reflected in state structures?  The analyst should keep in mind that even if the political authorities are responsive to popular will, the state may still be prone to informal exclusion if dominant norms of intolerance are in place.  The question goes beyond the formal institutional structures to the belief system underpinning them.  The indicators of tolerance should be evident in the extent of effort expended in order not to exclude certain groups from participation in the political process.  In certain countries, polling data on views toward other groups may be available; if reliable, such data may provide important clues as to the prevalent norms.

**C1c. Cohesion among ruling elites.** Is there prevailing agreement among the ruling elites about upholding the existing political structures?  Are there certain elites whose actions demonstrate ambiguous views about the existing political structures?  The specific indicators of cohesion should be evident from internal debates among the ruling elites.  Public actions, such as obvious distancing by some of the elites, may also provide clues.

**C1d. Composite assessment of accommodative capability.** The assessment of the accommodative capability of the state is a composite picture based on C1a–C1c, giving most importance to factors of responsiveness (C1a), determined in C1a3.  Democratic states are bound to score better than authoritarian ones in responsiveness.  However, responsiveness may vary considerably among democratic states.  States with new democratic structures may be quite different in their level of responsiveness from longstanding states where inclusionary institutions have had a chance to develop and norms of tolerance have been internalized.  In this sense, C1b can be an important modifier of the assessment arrived at in C1a3.  In addition, the existence of factions in the ruling elite that have doubts about upholding the existing political structures lowers the accommodative capability of the state, even if the assessments in C1a and C1b may give the state an otherwise high mark in this category.

## C2. Fiscal and Economic Capability

Three sets of questions address the issue of economic resources that the state can amass in its competition with the mobilized group.  The category is important because it establishes the bounds on how well the state can deal with the group and its demands without resorting to force. The first set of questions deals with the state's overall fiscal health (questions coded C2a). These questions address the financial resources readily available to the state. The second set of questions pertains to the resources that potentially could be extracted from the society by the state machinery and made available for the competition with the mobilized group (questions coded C2b). The third set of questions deals with the size and wealth of the main group that is represented by the ruling elites (questions coded C2c). On the basis of the assessments in each category, an overall assessment of the state's fiscal and economic capabilities needs to be made (guidelines coded C2d).

**C2a. Fiscal health.** Is the state engaged in deficit spending, or is it amassing surpluses? If it is deficit spending, has the trend gone on for some time? What is the burden imposed on the state treasury by payment of debt and interest? How is the budget appropriated, and could the state make substantial shifts in how the funds are appropriated? For example, could substantial funds be transferred from defense to specifically targeted social spending? Data to determine the health of the treasury are generally available from official sources, though they should be assessed first for reliability.

**C2b. Resource extraction potential.** How much slack is there in terms of the state being able to increase the rate of extraction of resources from the society? Is the tax burden at such a high level that a significant increase will be counterproductive (flight of capital, substantial erosion of political support)? Or are tax burdens fairly low and are there readily identifiable areas of wealth? And is the bureaucratic machinery in place able to collect the additional resources? Some of the data to answer these questions should be readily available in official publications. The comparative assessment of whether the tax burden is low or high needs to be made on the basis of real tax rates per capita and investment climate (as rated by international banks) against countries at similar development levels (potential competitors for foreign investment). An evaluation of the efficiency of the tax collection bureaucracy (based on indicators such as estimates of levels of tax evasion) also needs to be made.

**C2c. Size and wealth of ruling elites.** What is the size and level of wealth of the ruling elite's core constituency (the group whose interests the ruling elite represents the most)? What are the comparative wealth levels of this group vis-à-vis the population as a whole? How high of a real tax burden (with taxes understood in the wide meaning of the term) is the group facing at this time? And what are the levels of tax burden that the core group might be willing to endure in order to support the ruling elite? In other words, is it likely that the main social base for the ruling elite will be willing to endure higher burdens than the general population in order to keep its privileged position? Data to answer these questions stem more likely than not from estimates derived from general tax burden levels.

**C2d. Composite assessment of fiscal and economic capability.** The assessment of the fiscal and economic capability of the state is based on a composite of C2a–C2c, so as to incorporate both the present and the potential level of mobilization of resources by the state. All three factors play a major role, and strength in one area can offset weaknesses in another.

## C3. Coercive Capability

Four separate sets of questions address the ability of the state to marshal and use force in its competition with the mobilized group. First, what are the structural command and control links between the ruling elites and the apparati of violence, such as the police, militia, and the armed forces (questions coded C3a)? This is the most important of the four factors. Second, what is the composition—at the rank-and-file level—of these apparati (questions coded C3b)? Third, what are the norms toward and experience with respect to the use of force against internal opponents (questions coded C3c)? Fourth, how capable are the apparati of violence in dealing with internal opposition (questions coded C3d)? A final assessment of the coercive capability is based on an examination of all these factors (guidelines coded C3e).

**C3a. Command and control over the apparati of violence.** What are the mechanisms for ordering the use of the police and/or the military against internal opponents? Is the process subject to the influence and opinions of others besides just the top executive authority? Do any other political structures—besides the executive—affect the extent of employment of the apparati of violence against domestic opponents?  The indicators of the structural constraints (or lack thereof) on executive authority and command of use of the apparati of violence internally are embedded in the legal rules governing the use of force.  An evaluation of how closely such rules are followed in practice is an essential component of the assessment.

**C3b.  Composition of the apparati of violence.**  Who serves in the apparati of violence?  Are some groups overrepresented?  Are the rank and file in the apparati of violence likely to follow the orders of the top authorities?  How likely are the rank and file to identify with an aggrieved group?  The data on the composition of the police and the military may be sensitive, especially in states where internal problems may be severe.  Intelligence sources may be needed to supplement data from open sources.  The assessment should focus on the units most likely to be used for internal purposes, though data for all of the police and the military should be taken into account.

**C3c.  Norms toward domestic use of violence.**  What is the state's propensity for the use of force, based on historical tradition and experiences?  Is there a longstanding reputation of the state to rely on the apparati of violence to quell internal dissent?  Have the apparati of violence been used for such purposes within the past two to three decades (one generation)?  The data for this assessment are openly available and based on the historical record.

**C3d.  Suitability of the force for domestic use.**  Are the equipment and training of the apparati of violence suitable for quelling internal unrest?  Are there units (in either the police or the military) that have a specific internal orientation (such as anti-riot units)?  How large are such forces?  The data for this assessment should be openly available, though some details (equipment of the anti-riot police) may be more difficult to locate.  Evaluations of the training of domestically tasked units by outside observers also would be helpful in assessing the proficiency of such formations.

**C3e.  Composite assessment of coercive capability.**  The assessment of the coercive capability of the state is based on a composite of C3a–C3d, with particular attention to C3a (because of its crucial role in constraining the decisionmaking process on the use of force).

## D. STRATEGIC BARGAINING

The bargaining process between a mobilized group and the state determines whether the outcome of the process will be violence, peaceful reconciliation, or some point in between. Both the group's and the state's preferences are structured by their own strengths and weaknesses as well as the opponent's strengths and weaknesses. The optimal strategy is one based on maximum cost-effectiveness (in political terms). In other words, the strategies the group and the state will pursue stem from the specific characteristics of each relative to the other. Rather than questions for the analyst, this section consists of guidelines, for the task here is to model the interaction on the basis of the data gathered so far.

Modeling the interactive process to anticipate the potential for violence requires placing the information gathered so far on the group and the state into a format that categorizes each into a specific state or group type. Using the two tables that describe the state and group capacities (guidelines for measuring the capacities are coded D1), a composite picture emerges of the essential strengths and weaknesses of the group and the state. The interaction between each type of group and each type of state is then simulated with the help of two matrices (guidelines on using the matrices are coded D2). A comparison of the preferences of the state and the group illustrates a likely outcome. Whether the outcome is violent or not addresses the fundamental point of the entire exercise—that is, the likelihood of violence.

The following outline presents the organization of the guidelines and the logical connections between the steps that an analyst needs to perform in order to come up with a final assessment on the likelihood of strife that might result from the bargaining process. The actual guidelines follow the outline.

**Outline of Guidelines for Assessing:**
**D. Strategic Bargaining**

**D1. MEASURING THE CAPACITIES OF THE GROUP AND THE STATE**

    **D1a.** Measuring the group's capacities

        ***D1a1.*** *Group capacity: leadership*

        ***D1a2.*** *Group capacity: sustainment*

        ***D1a3.*** *Group capacity: cohesiveness*

        ***D1a4.*** *Locating the group type*

    **D1b. Measuring the state's capacities**

        ***D1b1.*** *State capacity: leadership*

        ***D1b2.*** *State capacity: fiscal position*

        ***D1b3.*** *State capacity: regime type*

        ***D1b4.*** *Locating the state type*

**D2. COMPARING PREFERENCES WITHIN THE BARGAINING PROCESS**

    **D2a.** Preferences of the group toward the state

    **D2b.** Preferences of the state toward the group

    **D2c.** Final outcome: preferences for violence

## D1.  Measuring the Capacities of the Group and the State

The first step is to measure the capacities of the group (guidelines coded D1a) and the state (guidelines coded D1b) in three critical areas. The critical areas are similar, though not the same, for the state and the group. The data for the measurement are based on assessments from previous sections (coded B and C). The measurements categorize the group and the state into one of eight specific types. The eight types comprise all of the possible combinations of the three categories, each coded in a binary fashion. The analyst must make some difficult choices at this point. The gathered data can structure the decision, but there is no way to avoid subjective reasoning in making an assessment based on only two choices.

**D1a. Measuring the group's capacities.** Categorizing the group in terms of three essential capacities—accommodative, sustainment, and cohesiveness—requires a binary assessment of leadership, resources, and the level of popular support of the group. Accommodative capacity measures the group's ability to accommodate its goals to the other competing social formations, especially the state. Sustainment capacity measures the group's ability to sustain its political campaign as it attempts to redress its grievances. Cohesiveness measures the group's ability to maintain group identity in the process of pursuing its group aims. The manner and strength of the group's mobilization determines its capacity in each of the three critical categories. Arriving at the determination means ranking the group in three areas: leadership (guidelines coded D1a1), sustainment (guidelines coded D1a2), and cohesiveness (guidelines coded D1a3). The specific rankings then determine the group type (guidelines coded D1a4).

*D1a1. Group capacity: leadership.* The analyst must rate the group in terms of leadership, using the choices of either strong or weak. Strong leadership is defined as having the following characteristics: (1) self-confident and secure in its position, (2) willing to take some risks of losing popularity in order to achieve its ends, and (3) owning a clear view of the goals it wants to accomplish. Weak leadership is defined as having the following characteristics: (1) conscious of other potential leaders, (2) cannot afford to take risks and obliged to appeal to the group broadly, even if such action compromises its goals, and (3) wavers in dedication to the goals. The assessments derived in the section on group mobilization regarding leadership (questions coded B3) should form the building block for a rating of the leadership.

*D1a2. Group capacity: sustainment.* The analyst must rate the group in terms of resource support, using the choices of either good or weak. Good resource support is defined as: (1) sufficient support to meet all near-term objectives, most mid-term objectives, and some long-term objectives, (2) real prospects of increased support as the group gains momentum, and (3) available support is suited to means and ends. Weak resource support is defined as: (1) insufficient support to meet even all near-term objectives, causing expenditure of considerable efforts in obtaining support, perhaps even driving the group into survival mode, (2) limited or uncertain prospects of increased support, and (3) mismatch between available support and means and ends. The assessments derived in the section on group mobilization regarding resources and organization (questions coded B4) as well as foreign support (questions coded B5) should form the building block for a rating of resource support.

*D1a3. Group capacity: cohesiveness.* The analyst must rate the group in terms of popular support, using the choices of either broad or weak. Broad popular support is defined as: (1) strong appeal of the ideas to the group and (2) potential for sympathy and/or support from other groups. Weak popular support is defined as: (1) weak resonance of the ideas beyond the leadership (ideas are either not fully or not widely accepted) and (2) no potential for support from other groups. The assessments derived in the section on group mobilization regarding incipient changes (questions coded B1), galvanizing events (questions coded B2), and the overall strength of mobilization and group identity (questions coded B6) should form the building blocks for a rating of popular support.

*D1a4. Locating the group type.* Based on a rating in each of the three categories, the analyst should locate the group type for the specific group being examined. Each type of a group (one of eight types, ranging from A to H), has corresponding capacities in three critical areas (see Table A.1).

**D1b. Measuring the state's capacities.** Categorizing the state in terms of three essential capacities—accommodative, sustainment, and coercive—requires a binary assessment of leadership, the fiscal health of the state, and the regime type. Accommodative capacity measures the state's ability to accommodate the group's demands. Sustainment capacity measures the state's ability to sustain its political preferences in competition with the group. Coercive capacity measures the state's ability to coerce the opponents into compliance. The political structure of the state, the state's potential for the use of force, and its fiscal and economic health determine its capacity in each of the three critical categories. Arriving at the determination means ranking the state in three areas: leadership (guidelines coded D1b1), fiscal health (guidelines coded D1b2), and regime type (guidelines coded D1b3). The specific rankings then determine the state type (guidelines coded D1b4).

Table A.1

Capacity of a Mobilized Group

| TYPE OF MOBILIZED GROUP | | CAPACITY | | |
|---|---|---|---|---|
| Code | Descriptors | Accommodative | Sustainment | Cohesiveness |
| A | Strong leadership<br>Good resource support<br>Broad popular support | High | High | High |
| B | Weak leadership<br>Good resource support<br>Broad popular support | Low | High | Low |
| C | Strong leadership<br>Weak resource support<br>Broad popular support | High | Low | High |
| D | Strong leadership<br>Good resource support<br>Weak popular support | Low | High | High |
| E | Weak leadership<br>Weak resource support<br>Broad popular support | Low | Low | Low |
| F | Weak leadership<br>Weak resource support<br>Weak popular support | High | Low | Low |
| G | Strong leadership<br>Weak resource support<br>Weak popular support | Low | Low | Low |
| H | Weak leadership<br>Good resource support<br>Weak popular support | Low | High | Low |

*D1b1. State capacity:  leadership.* The analyst must rate the state in terms of leadership, using the choices of either strong or weak.  Strong leadership is defined in the same manner as for the group, and it has the following characteristics:  (1) self-confident and secure in its position, (2) willing to risk losing popularity in order to achieve its ends, and (3) owning a clear view of the goals it wants to accomplish.  Weak leadership is also defined in the same manner as the group, and it has the following characteristics:  (1) conscious of other potential leaders, (2) cannot afford to take risks and obliged to appeal to the group broadly, even if such action compromises its goals, and (3) wavers in dedication to the goals. The assessments derived in the section on the political structure (questions coded C1) are the shaping influences on any specific leadership in power. The structural elements are more important than the specific personalities in question, though some knowledge of the specific leadership is also necessary.

*D1b2. State capacity: fiscal position.* The analyst must rate the state in terms of its fiscal position, using the choices of either strong or weak. Strong fiscal position is defined as having the following characteristics: (1) budget surpluses or low deficits, (2) room for major reallocations within the structure of the budget, (3) enough wealth for increased revenue generation. Weak fiscal position is defined as having the following characteristics: (1) deep and/or prolonged deficit spending, (2) highly constrained room for reallocation within the budget, and (3) limited potential for increased revenue generation due to limited wealth. The assessments derived in the section on the fiscal and economic health of the state (questions coded C2) should form the building blocks for a rating of the fiscal position.

*D1b3. State capacity: regime type.* The analyst must rate the state by regime type, using the choices of either inclusive or exclusive. While the two choices have some overlap with democratic and nondemocratic regime types, the terms are not the same, for they focus both on the normative as well as institutional aspects. An inclusive regime is defined as having the following characteristics: (1) competitive elections and unrestricted rights to assembly, (2) uncensored media, (3) limits on the executive power, and (4) prevailing norms of tolerance. An exclusive regime is defined as having the following characteristics: (1) lack of competitive elections and/or restrictions on rights to assembly, (2) restrictions on or censorship of the media, (3) no real check on the executive power, and (4) norms of tolerance are either weak or nonexistent. Almost all democratic regimes are inclusive, but not all inclusive regimes are democratic. Similarly, almost all oligarchic or authoritarian regimes are exclusive, but not all exclusive regimes are nondemocratic. The assessments on political structure (questions coded C1) and the apparati of coercion (questions coded C3) should form the building blocks for rating a state on regime type.

## Table A.2

## Capacity of the State

| TYPE OF STATE | | CAPACITY | | |
|---|---|---|---|---|
| Code | Descriptors | Accommodative | Sustainment | Coercive |
| A | Strong leadership<br>Strong fiscal position<br>Inclusive regime | High | High | Low |
| B | Weak leadership<br>Strong fiscal position<br>Inclusive regime | Low | Low | High |
| C | Strong leadership<br>Weak fiscal position<br>Inclusive regime | High | Low | Low |
| D | Strong leadership<br>Strong fiscal position<br>Exclusive regime | Low | Low | High |
| E | Weak leadership<br>Weak fiscal position<br>Inclusive regime | High | Low | Low |
| F | Weak leadership<br>Weak fiscal position<br>Exclusive regime | Low | Low | Low |
| G | Strong leadership<br>Weak fiscal position<br>Exclusive regime | Low | Low | High |
| H | Weak leadership<br>Strong fiscal position<br>Exclusive regime | Low | High | High |

*D1b4. Locating the state type.*  Based on a rating in each of the three categories, the analyst can locate the state type for the specific state being examined.  Each type of a state (one of eight types, ranging from A to H), has corresponding capacities in three critical areas (see Table A.2).

## D2. Comparing Preferences Within the Bargaining Process

Once the analyst has defined the group and the state as fitting one particular state and group type, the next step is to locate the specific preferences of each on a matrix of preference choices. This is essentially a mechanical step, carried out in two stages: figuring out the preferences of the group toward the state (guidelines coded D2a) and figuring out the preferences of the state toward the group (guidelines coded D2b). The interaction of the two preferences makes up the final assessment of whether violence is likely (guidelines coded D2c).

**D2a. Preferences of the group toward the state.** Using Matrix A.1, the analyst should locate the row appropriate to the specific type of state that the group is facing. For example, if the specific determination from D1 has concluded that the situation pits group type D against a state type C, the analyst would locate group D in the column at the left of the matrix and, using the row of group type D, look at the set of three preferences in the column under state type C. Circling the specific three preferences in the matrix, the choices are: (1) Neg, (2) Exp, and (3) Int. (An explanation of the preferences follows in D2c.)

**D2b. Preferences of the state toward the group.** Using Matrix A.2, the analyst should locate the column appropriate to the specific type of group that the state is facing. As in the example used in D2a, if the specific determination from D1 has concluded that the situation pits state type C against a group type D, the analyst would locate state type C in the row at the top of the matrix and, using the row in group type D, look at the set of three preferences in the column under state type C. Circling the specific three preferences in the matrix, the choices are: (1) Rep, (2) Exp, and (3) Neg. (An explanation of the preferences follows in D2c.)

**D2c. Final outcome: preferences for violence.** Combining D2a and D2b determines the likelihood of violence in the given case. There are four preference choices used in the matrices. For the state, the preferences are: "Rep" (repress), meaning the use of violence against the group; "Exp" (exploit), meaning the use of nonviolent means to compete with the group; "Neg" (negotiate), meaning the use of peaceful negotiation; and "Sur" (surrender), meaning the state surrenders to the demands of the group. For the group, the preferences

## Matrix A.1

### Mobilized Group Preferences

| Mobilized Group Type (Table A.1) | State Type (Table A.2) | | | | | | | |
|---|---|---|---|---|---|---|---|---|
| | A | B | C | D | E | F | G | H |
| A | 1. neg<br>2. exp<br>3. int | 1. exp<br>2. int<br>3. neg | 1. neg<br>2. exp<br>3. int | 1. int<br>2. exp<br>3. neg | 1. neg<br>2. exp<br>3. int | 1. exp<br>2. int<br>3. neg | 1. int<br>2. exp<br>3. neg | 1. exp<br>2. int<br>3. neg |
| B | 1. exp<br>2. int<br>3. neg | 1. int<br>2. exp<br>3. neg | 1. neg<br>2. exp<br>3. int | 1. int<br>2. neg<br>3. int | 1. int<br>2. exp<br>3. neg | 1. int<br>2. exp<br>3. neg | 1. exp<br>2. neg<br>3. int | 1. exp<br>2. int<br>3. neg |
| C | 1. neg<br>2. exp<br>3. int | 1. exp<br>2. neg<br>3. int | 1. neg<br>2. exp<br>3. int | 1. int<br>2. neg<br>3. exp | 1. exp<br>2. int<br>3. neg | 1. int<br>2. exp<br>3. neg | 1. int<br>2. exp<br>3. neg | 1. exp<br>2. int<br>3. neg |
| D | 1. neg<br>2. exp<br>3. int | 1. exp<br>2. neg<br>3. int | 1. neg<br>2. exp<br>3. int | 1. int<br>2. exp<br>3. neg | 1. int<br>2. exp<br>3. int | 1. int<br>2. exp<br>3. neg | 1. exp<br>2. neg<br>3. int | 1. exp<br>2. neg<br>3. int |
| E | 1. exp<br>2. neg<br>3. int | 1. exp<br>2. int<br>3. neg | 1. exp<br>2. neg<br>3. int | 1. exp<br>2. neg<br>3. int | 1. exp<br>2. int<br>3. neg | 1. int<br>2. exp<br>3. neg | 1. neg<br>2. exp<br>3. sur | 1. exp<br>2. int<br>3. neg |
| F | 1. neg<br>2. exp<br>3. sur | 1. exp<br>2. neg<br>3. sur | 1. neg<br>2. exp<br>3. int | 1. neg<br>2. exp<br>3. sur | 1. exp<br>2. int<br>3. neg | 1. exp<br>2. int<br>3. neg | 1. exp<br>2. neg<br>3. sur | 1. exp<br>2. int<br>3. neg |
| G | 1. exp<br>2. neg<br>3. int | 1. int<br>2. exp<br>3. neg | 1. neg<br>2. exp<br>3. int | 1. int<br>2. neg<br>3. exp | 1. int<br>2. exp<br>3. neg | 1. int<br>2. exp<br>3. neg | 1. exp<br>2. int<br>3. neg | 1. int<br>2. exp<br>3. neg |
| H | 1. exp<br>2. neg<br>3. sur | 1. exp<br>2. int<br>3. neg | 1. exp<br>2. int<br>3. neg | 1. exp<br>2. int<br>3. neg | 1. int<br>2. exp<br>3. neg | 1. int<br>2. exp<br>3. neg | 1. exp<br>2. int<br>3. neg | 1. exp<br>2. neg<br>3. int |

neg = negotiation; exp = exploitation; int = intimidation; sur = surrender

## Matrix A.2

### State Preferences

| Mobilized Group Type (Table A.1) | State Type (Table A.2) | | | | | | | |
|---|---|---|---|---|---|---|---|---|
| | A | B | C | D | E | F | G | H |
| A | 1. neg<br>2. exp<br>3. rep | 1. neg<br>2. exp<br>3. rep | 1. neg<br>2. exp<br>3. rep | 1. rep<br>2. neg<br>3. rep | 1. neg<br>2. exp<br>3. sur | 1. rep<br>2. exp<br>3. neg | 1. rep<br>2. exp<br>3. neg | 1. exp<br>2. rep<br>3. neg |
| B | 1. exp<br>2. neg<br>3. rep | 1. neg<br>2. exp<br>3. rep | 1. neg<br>2. exp<br>3. rep | 1. exp<br>2. rep<br>3. neg | 1. neg<br>2. exp<br>3. rep | 1. exp<br>2. rep<br>3. neg | 1. exp<br>2. neg<br>3. rep | 1. exp<br>2. rep<br>3. neg |
| C | 1. neg<br>2. exp<br>3. rep | 1. neg<br>2. exp<br>3. rep | 1. neg<br>2. exp<br>3. rep | 1. rep<br>2. exp<br>3. neg | 1. neg<br>2. exp<br>3. sur | 1. rep<br>2. exp<br>3. neg | 1. rep<br>2. exp<br>3. neg | 1. rep<br>2. exp<br>3. neg |
| D | 1. rep<br>2. neg<br>3. exp | 1. neg<br>2. rep<br>3. exp | 1. rep<br>2. exp<br>3. neg | 1. rep<br>2. exp<br>3. neg | 1. neg<br>2. exp<br>3. rep | 1. rep<br>2. exp<br>3. neg | 1. rep<br>2. exp<br>3. neg | 1. rep<br>2. exp<br>3. neg |
| E | 1. exp<br>2. neg<br>3. rep | 1. neg<br>2. exp<br>3. rep | 1. exp<br>2. neg<br>3. rep | 1. exp<br>2. neg<br>3. rep | 1. neg<br>2. exp<br>3. rep | 1. rep<br>2. exp<br>3. neg | 1. exp<br>2. neg<br>3. rep | 1. rep<br>2. exp<br>3. neg |
| F | 1. rep<br>2. exp<br>3. neg | 1. neg<br>2. exp<br>3. rep | 1. rep<br>2. exp<br>3. neg | 1. rep<br>2. exp<br>3. neg | 1. neg<br>2. rep<br>3. exp | 1. exp<br>2. rep<br>3. neg | 1. rep<br>2. exp<br>3. neg | 1. rep<br>2. exp<br>3. neg |
| G | 1. rep<br>2. neg<br>3. exp | 1. neg<br>2. exp<br>3. rep | 1. rep<br>2. exp<br>3. neg | 1. rep<br>2. exp<br>3. neg | 1. neg<br>2. exp<br>3. rep | 1. rep<br>2. exp<br>3. neg | 1. rep<br>2. exp<br>3. neg | 1. exp<br>2. rep<br>3. neg |
| H | 1. rep<br>2. exp<br>3. neg | 1. neg<br>2. exp<br>3. rep | 1. rep<br>2. exp<br>3. neg | 1. rep<br>2. exp<br>3. neg | 1. neg<br>2. exp<br>3. rep | 1. exp<br>2. rep<br>3. neg | 1. rep<br>2. exp<br>3. neg | 1. exp<br>2. rep<br>3. neg |

neg = negotiation; exp = exploitation; rep = repression; sur = surrender

are: "Int" (intimidate), meaning the use of violence against the state; "Exp" (exploit), meaning the use of nonviolent means to compete with the state; "Neg" (negotiate), meaning the use of peaceful negotiation; and "Sur" (surrender), meaning the group surrenders to the pressure by the state.

# ABOUT THE AUTHORS

**Daniel L. Byman** is a policy analyst at RAND; his work focuses on civil wars, the Middle East, and other topics related to U.S. national security.

**Graham E. Fuller** is a consultant at RAND, and a former National Intelligence Officer for the Middle East at CIA.

**Sandra F. Joireman** is an Associate Professor of Political Science at Wheaton College in Wheaton., Illinois and author of *Property Rights and Political Development in Ethiopia and Eritrea* (Athens, Ohio: Ohio University Press, 2000).

**Pearl-Alice Marsh,** Ph.D. is a political scientist and a former Senior Research Associate at the Joint Center for Political and Economic Studies.

**Thomas S. Szayna** is a political scientist at RAND; his research interests include contemporary security environment, intrastate conflict, and peacekeeping.

**Ashley J. Tellis** is a senior policy analyst at RAND. His work focuses on international relations theory, military strategy and proliferation issues, South Asian politics, and U.S.-Asian security relations.

**James A. Winnefeld** is a retired RAND senior analyst and a former naval aviator.

**Michele Zanini** is a Ph.D. candidate at the RAND Graduate School and an author of a number of RAND publications on NATO strategy in the Balkans and Mediterranean, terrorism, and European defense planning.